W0228142

Functional Studies Using NMR

Edited by

V. Ralph McCready, Martin Leach and Peter J. Ell

With 146 Figures

Springer-Verlag
London Berlin Heidelberg New York
Paris Tokyo

V. Ralph McCready, MSc, MRCP, FRCR
Consultant in Nuclear Medicine, Ultrasound and NMR, Royal Marsden
Hospital, Downs Road, Sutton, Surrey, UK

Martin Leach, MSc, PhD
Lecturer in Medical Physics (NMR), Joint Department of Physics,
Institute of Cancer Research and Royal Marsden Hospital, Downs
Road, Sutton, Surrey, UK

Peter J. Ell, MD, MSc, PD, MRCP, FRCR
Director, Institute of Nuclear Medicine, The Middlesex Hospital
Medical School, Mortimer Street, London, UK

ISBN-13: 978-1-4471-1412-3   e-ISBN-13: 978-1-4471-1410-9
DOI: 10.1007/978-1-4471-1410-9

Library of Congress Cataloging-in-Publication Data
Functional Studies using NMR.
"This volume is based on a series of lectures delivered at a one-day teaching symposium on
functional and metabolic aspects of NMR measurements held at the Middlesex Hospital
Medical School on September 1st 1985 as part of the European Nuclear Medicine Society
congress" – Pref.
Includes bibliographies and index.
1. Nuclear magnetic resonance spectroscopy – Congresses. 2. Magnetic resonance imaging
– Congresses. 3. Metabolism – Congresses. I. McCready, V.R. II. Leach, M.O. (Martin
Osmund), 1952–   . III. Ell, Peter Josef. IV. European Nuclear Medicine Society. [DNLM:
1. Metabolism – congresses. 2. Nuclear Magnetic Resonance – congresses. QU 25 F979
1985] QP519.9.N83F86   1986   616.07'57 86-22121

The work is subject to copyright. All rights are reserved, whether the whole or part of the
material is concerned, specifically those of translation, reprinting, re-use of illustrations,
broadcasting, reproduction by photocopying, machine or similar means, and storage in
data banks.

© Springer-Verlag Berlin Heidelberg 1987
Softcover reprint of the hardcover 1st edition 1987

The use of general descriptive names, trade marks, etc. in this publication, even if the
former are not to be taken as a sign that such names, as understood by the Trade Marks and
Merchandise Marks Act, may accordingly be used freely by anyone.

Product Liability: The publisher can give no guarantee for information about drug dosage
and application thereof contained in this book. In every individual case the respective user
must check its accuracy by consulting other pharmaceutical literature.

Filmset by Tradeset, Welwyn Garden City, Hertfordshire.

2128/3916-543210

Preface

This volume is based on a series of lectures delivered at a one-day teaching symposium on functional and metabolic aspects of NMR measurements held at the Middlesex Hospital Medical School on 1st September 1985 as a part of the European Nuclear Medicine Society Congress.

Currently the major emphasis in medical NMR in vivo is on its potential to image and display abnormalities in conventional radiological images, providing increased contrast between normal and abnormal tissue, improved definition of vasculature, and possibly an increased potential for differential diagnosis. Although these areas are undeniably of major importance, it is probable that NMR will continue to complement conventional measurement methods. The major potential benefits to be derived from in vivo NMR measurements are likely to arise from its use as an instrument for functional and metabolic studies in both clinical research and in the everyday management of patients. It is to this area that this volume is directed.

Although most of the areas of work described are as yet at an early stage of development, it is evident that NMR is a powerful and many faceted tool for investigating function and metabolism in vivo. It combines high quality anatomical imaging, at times surpassing the resolution of CT scanners, with the potential for using functional and specific agents in a manner akin to nuclear medicine whilst providing multiple parameters for studying tissue changes and for discriminating tissue type. NMR spectroscopy gives the potential to study cellular metabolism in localised regions of tissue. Functional NMR imaging methods allow the measurement of blood flow, currently providing results similar to Doppler ultrasound, but offering the possibility of being extended to measure organ perfusion. It is also possible to monitor organ function with high resolution using NMR, for example, in cardiac function measurements.

This book is intended to provide the background information required for the reader who is not familiar with NMR, whilst also providing an up-to-date summary of advances and current work in the areas of interest in metabolic, physiological and functional studies.

We are particularly grateful to the contributing authors who have given such an exciting and comprehensive treatment to the different areas covered by this volume and who ensured that the symposium was a considerable success.

Sutton, England   V. Ralph McCready
June 1986   Martin Leach
                                                                  Peter J. Ell

Contents

Contributors

R. Bachus, PhD
Research Scientist, Siemens AG., Medical Division, Henkestr. 127,
D-8520 Erlangen, West Germany

G. Bielke, PhD
NMR Department, Deutsche Klinik fur Diagnostik, Aukammallee 33,
D-6200 Wiesbaden, West Germany

G. Braeckle, PhD
Research Scientist, Siemens AG., Medical Division, Henkestr. 127,
D-8520 Erlangen, West Germany

G.M. Bydder, MRCP
Senior Lecturer, Royal Postgraduate Medical School, University of
London, Hammersmith Hospital, Ducane Road, London, UK

R.E. Coupland, MD, PhD, DSc
Professor of Human Morphology, Department of Human Morphology,
University of Nottingham Medical School, Clifton Boulevard,
Nottingham, UK

Margaret A. Foster, BSc, MSc, PhD
Senior Lecturer, Department of Bio-Medical Physics and
Bio-Engineering, University of Aberdeen, Foresterhill, Aberdeen, UK

L.D. Hall, BSc, PhD, FRSC, FRS(Can)
Herchel Smith Professor of Medicinal Chemistry, Laboratory for
Medicinal Chemistry, University of Cambridge School of Clinical
Medicine, Level 4, Radiotherapeutic Centre, Addenbrooke's Hospital,
Hills Road, Cambridge, UK

H.P. Higer, MD
NMR Department, Deutsche Klinik fur Diagnostik, Aukammallee 33,
D-6200 Wiesbaden, West Germany

P.G. Hogan, BSc
Laboratory for Medicinal Chemistry, University of Cambridge School of
Clinical Medicine, Level 4, Radiotherapeutic Centre, Addenbrooke's
Hospital, Hills Road, Cambridge, UK

A. Everette James Jr., MD
Professor of Radiology and Medical Administration, Chairman,
Department of Radiology and Radiological Sciences, Vanderbilt
University Medical Center, Nashville, Tennessee, 37232, USA

H. Koenig, PhD
Research Scientist, Siemens AG., Medical Division, Henkestr. 127,
D-8520 Erlangen, West Germany

M.V. Kulkarni, MD
Associate Professor of Radiology, Director, Magnetic Resonance
Imaging, University of Texas, Health Sciences Center at Houston, 6431
Fannin 2.124 M.S.B. Houston, Texas, 77030, USA

M.O. Leach, PhD
Lecturer in Medical Physics (NMR), Department of Physics, Royal
Marsden Hospital, Downs Road, Sutton, Surrey, UK

P. Mansfield, BSc, PhD,
Professor of Physics, Department of Physics, University of Nottingham,
University Park, Nottingham, UK

V. Ralph McCready, MRCP, FRCP
Consultant in Charge, Department of Nuclear Medicine, Royal Marsden
Hospital, Downs Road, Sutton, Surrey, UK

S. Meindl, Dipl. Phys.
NMR Department, Deutsche Klinik fur Diagnostik, Aukammallee 33,
D-6200 Wiesbaden, West Germany

E. Mueller, PhD
Research Scientist, Siemens AG., Medical Division, Henkestr. 127,
D-8520 Erlangen, West Germany

H.P. Niendorf, MD
Schering, Postfach 65 03 11, D-1000 Berlin 65, West Germany

C. Leon Partain, MD, PhD
Professor, Radiology and Biomedical Engineering, Director, Nuclear
Medicine and Magnetic Resonance Imaging, Vanderbilt University
Medical Center, Nashville, Tennessee, 37232, USA

J.A. Patton, PhD
Associate Professor of Radiology, Associate Director, Nuclear Medicine
and Magnetic Resonance Imaging, Vanderbilt University Medical
Center, Nashville, Tennessee, 37232, USA

P. Pfannenstiel, MD
Professor of Internal Medicine and Nuclear Medicine, Head of
Departments of Nuclear Medicine and NMR Tomography, Deutsche
Klinik fur Diagnostik, Aukammallee 33, D-6200 Wiesbaden, West
Germany

E.R. Reinhardt, PhD
Director, Research Department, Siemens AG., Medical Division,
Henkestr. 127, D-8520 Erlangen, West Germany

P. Sandler, MD
Associate Professor, Radiology and Medicine, Clinical Director, Nuclear
Medicine, Vanderbilt University Medical Center, Nashville, Tennessee,
37232, USA

A.N. Stevens, BSc, PhD
MR Specialist, IGE. Medical Systems, Coolidge House, 260 Bath Road,
Slough, Berks, UK

S. Richard Underwood, MA, MRCP
Senior Clinical Research Fellow, Magnetic Resonance Unit, The
National Heart and Chest Hospitals, 30 Britten Street, London, UK

W. von Seelen, PhD
Institute of Biophysics University, D-6500 Mainz, West Germany

H. Weber, PhD
Research Scientist, Siemens AG., Medical Division, Henkestr. 127,
D-8520 Erlangen, West Germany

I.R. Young, PhD
Hirst Research, P.O. Box 2, East Lane, Wembley, UK

1. The Relationship Between NMR and Other Functional Measurement Techniques

V. Ralph McCready

Introduction

Although nowadays radioisotope tests are synonymous with imaging, the original application of radioisotope studies was in the sphere of functional measurement. One of the earliest techniques for measuring in vivo metabolism was the use of radioiodine to study the function of the thyroid gland in health and disease (Rall 1956). The development of improved Geiger counting systems and scintillation detectors gave clinicians the opportunity to measure and image various aspects of body physiology including cerebral perfusion and cardiac function. The scintillation counter is a highly sensitive device but requires suitable collimation to define an area of interest within the body. The development of collimators and dual crystal counting systems produced an isosensitive technique which could measure radioactivity in any part of the body without the problem of errors due to attenuation. However the difficulty of ensuring that all of the uptake came from the organ alone, without the complications of including radiation from surrounding tissues, continued until the advent of rectilinear scanning and the use of the Anger camera. The current combination of gamma camera and a digital computer is a powerful one, enabling accurate estimation of radiation emanating from a defined area within the body. Emission tomography has further improved measurement techniques so that radioactivity distributions in three dimensions can be defined and measured (Webb et al. 1985).

More recently alternative non-invasive methods for studying human physiology have become available including CT, digital subtraction imaging, ultrasound and now nuclear magnetic resonance — the subject of this book. Enthusiasm for the latest technique often exceeds the accuracy and logic of using it in a particular situation. It is therefore worth comparing the value of these various techniques for studying the commoner aspects of human physiology.

Ideally the choice of a particular technique should depend mainly upon its accuracy. However, lack of toxicity, morbidity or hazards from radiation or other physical phenomena have to be taken into consideration. The hazards from radiation and toxicity have to be balanced against the necessity of carrying out the test. In most clinical situations where a diagnosis is needed, the hazard related to the test is usually minimal in relation to the morbidity of the disease.

The aspects of physiology and tissue analysis most amenable to non-invasive measurement can be conveniently divided into three groups: blood flow, metabolism and tissue characterisation. These will be considered in turn.

Blood Flow

Non-invasive techniques for measuring superficial and deep blood flow include Doppler ultrasound, radionuclides, X-ray contrast studies and nuclear magnetic resonance.

Ultrasound

Doppler ultrasound relies upon the fact that the apparent frequency of a constant frequency source depends upon the motion of the source relative to the receiver. If the source is moving towards the receiver the apparent frequency increases and vice versa. In practice, the receiver and source are the same and it is the velocity of the reflector that is measured. In clinical work it is the red blood cells that act as the reflector and it is their movement that enables blood flow to be measured. The velocity v is related to the Doppler shift by the equation:

$$f = 2 (v \cos \theta) f_o / c \tag{1}$$

where v = blood velocity magnitude, θ = angle between the ultrasound beam and the moving red cells, f_o = the frequency of the transmitted ultrasound and c = the speed of sound propagation in blood.

The velocity measurement can be converted into a flow value provided the cross-section diameter and angle of the vessel relative to the transducer can be measured. This is done using the B or real time scan. The assessment of organ or tissue flow using Doppler relies upon finding a single supplying vessel. In the case of the kidney where the supplying vessel can be found easily there is little problem. In other situations, such as the brain, alternative methods are better.

Qualitative Doppler ultrasound blood flow measurements have been used for many years in obstetrics where the frequency shift is converted into an audio signal and a suitable vessel found by trial and error. With this technique it is possible to detect pulsation of the foetal heart and placental blood flow without the necessity of organ imaging. Similar techniques have been used in examining the carotid arteries, the deep venous system of the leg and now the abdomen (Taylor et al. 1985). More recently Doppler systems have been combined with imaging devices so that the area being examined could be identified more accurately (Wells and Skidmore 1985). Initially it was necessary to switch from the two-dimensional mode, which uses pulsed ultrasound, to the continuous signal mode for the Doppler measurement. Now it is possible to combine both techniques into one image and in addition use colour to demonstrate the direction of the blood flow (Switzer and Nanda 1985).

The accuracy of Doppler for measuring blood flow relies upon the precise determination of the cross-section of the blood vessel (using the diameter) under study and the accurate estimation of the velocity of the blood. The cross-sectional area (A) of the vessel is given by the equation:

$$A = d^2 / 4 \tag{2}$$

where A = area and d = diameter of the vessel.

This can be measured most accurately by a pulsed system viewing the vessel at right angles. The accuracy in this case is limited by the axial resolution of the equipment. At angles other than 90° (usually the case) the diameter will be underestimated because of the ultrasound beam width. The correct determination of θ (the angle between the beam and the vessel) is also crucial in determining the velocity and the area of the vessel — the largest errors being produced when the angle between vessel and beam is smallest. The angle is usually measured directly and automatically on the two-dimensional scan. Pulsation of the artery or the vein alters the cross-sectional area so it is not possible to measure the diameter and blood velocity simultaneously. However, experimental verification of Doppler blood flow measurements shows a high level of accuracy in expert hands. The regression lines of measured and true flow are between 0.95 and 1.06 (Gill 1985).

Advantages of Ultrasound Techniques

The advantages of using ultrasound techniques are: economy, the general availability, the ability to use them at the bedside in patients who may be very ill and the lack of any radiation hazard. The great disadvantages are the necessity for an acoustic window bone and air forming an impenetrable barrier. To measure organ blood flow it is essential to identify a single supplying vessel but in many cases this is not possible. For functional studies the limitations are the difficulties in quantitation outlined above. In clinical practice, however, a qualitative measurement is often sufficient.

Radioisotope Determination of Blood Flow

Dynamic studies have formed the most interesting part of nuclear medicine for many years. They can be divided into quantitative and qualitative techniques.

The qualitative method of studying blood flow merely involves the intravenous or intra-arterial injection of a suitable radionuclide or pharmaceutical such as pertechnetate or labelled red cells and an appropriate imaging technique to display the passage of the radioactivity in real time (Ennis and Dowsett 1983). In spite of the limited resolution of gamma ray imaging it is possible to display the peripheral veins (Coltart and Wraight 1985; Ahmad et al. 1979), the heart and the large vessels on the arterial side of the body. It is also possible to demonstrate the perfusion of the liver, spleen and kidneys in their normal or transplanted position. With modern imaging systems and high activities of short-lived radionuclides it is possible to see the superficial venous system with a high degree of precision and it is usually possible to demonstrate the presence and site of obstruction. On the arterial side it is more difficult to see abnormalities due to the loss of resolution at depth but in general it is possible to diagnose most aneurysms, obstructions and cardiac shunts.

Clearance Studies

The simplest quantitative technique for studying blood flow is the clearance method where a suitable radionuclide is injected and the change in radioactivity due to washout measured by a collimated scintillation detector gamma camera. In general the clearance is in proportion to the blood flow:

$$C_t = C_o e^{-kt} \tag{3}$$

where C_o = initial concentration. C_t = concentration at time t and k = clearance rate. The clearance can be converted to actual blood flow provided the partition coefficient of the tracer between the tissue under examination and the blood is known.

The technique is limited to skin and superficial lesions. The limitations of the method centre around the fact that only the blood flow at the point of injection is measured. It is also apparent that the injection itself must disturb the local environment and so alter the blood flow. In addition, in repeated studies it is impossible to be sure that the same area or volume is being studied.

Flow Dilution Method

In this technique a radiotracer is injected into a vessel and the dilution downstream measured over a period of time. The area under the time/activity curve calibrated for the detection system gives the total blood flow in the vessel according to the formula:

$$F = I / CT = I / A \tag{4}$$

where F = blood flow, I = injected radioactivity, A = area under the dilution curve, C = mean concentration in units/unit volume and T = the duration of passage of the radioactivity.

It is obvious that for this technique to be accurate a non-diffusable radiotracer must be used and also enough radioactivity to give high statistical accuracy. The method has been used with good results in measuring arterial flow and cardiac output.

Uptake Techniques

A variety of uptake techniques are available for the measurement of organ blood flow. The ideal tracer should be taken up completely on the first pass, the blood flows are then found by measuring the distribution of the radioactivity throughout the tissues of the body. The fractional uptake of the total radioactivity multiplied by the cardiac output gives the blood flow in volume per unit time.

The only radiotracer which can theoretically give 100% extraction and fixation of the first pass is macroaggregated albumin. However this has to be injected intra-arterially and is therefore invasive. Agents which can be given intravenously with high extraction rates include rubidium (Rb) and hexamethyl PAO (HM-PAO) (Nowotnik et al. 1985). Thallium has also been used in the assessment of myocardial perfusion but while it is possible to see variations in perfusion from area to area it is very difficult to measure flow in ml min^{-1} (Handler et al. 1985). Rubidium 82 has been used similarly (Mullani et al. 1983). The usual application of these and similar tracers has been for the measurement of cerebral blood flow (Holman and Magistretti 1982; Holman et al. 1984).

While there is little or no problem is assessing the blood flow qualitatively and displaying the results as an image, there are difficulties in the accurate assessment of the uptake quantitatively. These result from the statistics of radioactivity, problems in calibration and in the derivation of the correct attenuation coefficient.

Advantages of Radionuclide Methods

For the quantitative or semi-quantitative assessment of blood flow in the clinical situation, the radioisotope methods are still likely to be the most competitive with NMR flow measurements. They are relatively cheap and non-invasive and produce results without an exact knowledge of the anatomy of the vasculature (Bassingthwaite and Holloway 1976). The techniques also measure some aspects of nutritional flow without the complication of corrections for the blood which bypasses the tumour or organ through a-v shunts.

Blood Flow Determination by X-ray Methods

For many years arteriography or venography has been the technique of choice in measuring blood flow qualitatively. The method involves the use of an iodine loaded contrast medium and relies upon recording the absorption of the X-rays by it. Attempts to make the technique quantitative have not really been successful: the difficulties include the problem of accurate densitometry when there is a varying tissue background. Digital subtraction imaging has been used with more success. The process is similar to conventional angiography but uses a digital computer to record the images. The first image is taken before the injection of the contrast medium and is used as a mask for subtracting tissue background from subsequent images. The resulting images display only the density due to the contrast together with artefacts due to movement. Using techniques similar to those in nuclear medicine it is possible to outline regions of interest in specific areas in the image and to plot time/density curves (Saddekni et al. 1985). It is also possible to produce parametric images (Bursch and Heintzen 1985) and the techniques are ideal to show anatomy. Quantitation is difficult but relative changes in blood volume, such as in the calculation of cardiac output, can probably be carried out with more accuracy than in radioisotope studies due to the better spatial resolution. An advantage of this technique is the ability to carry out some studies with an intravenous injection but a disadvantage is the requirement for contrast media which have a small but well-documented hazard in clinical use and can also affect the vessel walls. Inherent in the techniques are the problems of movement between the first and subsequent exposures due to swallowing, respiration or peristalsis.

Blood Flow Measurement Using NMR

Qualitative Studies

The techniques for measuring blood flow using NMR are still at an early stage of development. Flowing blood has a variable appearance on the NMR image depending upon the velocity and direction of flow and the pulse sequence used. In general flowing blood has a dark appearance, i.e. low signal (Bradley and Waluch 1985). The obvious advantage of this phenomenon is the ability to differentiate between blood vessels, blood and other structures without need for contrast. For example it has been possible to demonstrate left atrial myxomas using a gated technique (Conces et al. 1985). The ability to differentiate between blood vessels, other structures and masses is especially valuable in the neck and mediastinum. In other parts of the body NMR has been used successfully to diagnose cardiac malformations (Fletcher et al. 1984), cardiac and paracardiac masses (Amparo et al. 1984) and has shown advantages over echocardiography due to the ability to have a three-dimensional view of the structures (Go et al. 1985). The thoracic aorta (Glazer et al. 1985c), aortic dissection (Amparo et al. 1984) and arterial atherosclerotic disease (Herfkens et al. 1983) have all been visualized without the need for contrast.

The advantages over conventional X-ray imaging are obvious: the resolution is similar to angiography and much better than that found in radionuclide studies. Radioiodinated contrast is not, of course, required but whether contrast has a role in NMR flow studies has yet to be determined. The value of NMR for functional measurements is not so clear. Cardiac studies have shown that NMR can produce ejection fractions with similar results and accuracy to those found with radionuclide techniques (Underwood et al. 1986). The problems of defining the exact volume of the intracardiac chambers, i.e. using tomographic slices and determining the precise edge, are common to both techniques but in principle the NMR method has the potential to give greater accuracy, does not require radioactivity and does not have the errors associated with radioactivity measurement. The correlation between NMR measurements and cardiac phantoms has been excellent (Rehr et al. 1985). Functional studies of the heart using NMR are discussed in detail in Chap. 6.

Quantitative Studies

While the advantages of NMR in being able to demonstrate blood flow in anatomical slices have been immediately obvious, its use for quantitative studies is still under development. The basic principle is simple (Mills et al. 1984b) but the fact that the signal varies from vessel to vessel and even within vessels demonstrates the complex physics underlying the analysis of what is actually happening (Bradley and Waluch 1985). In principle the technique of measuring blood flow is simple. A pulse of RF is used to rotate the spins through 90°, then a 180° pulse is given to rephase the spins. After a period an emitted signal or echo is recorded. While this process takes place blood both enters and leaves the volume being studied. It can be seen that the signal will be reduced by some irradiated protons leaving the volume prior to their spin echo being received. The situation is complicated by the sensitivity of the technique to the variation in velocity between the vessel wall and the centre of the lumen.

There is a simple relationship between "high velocity signal loss" and flow when the flow is perpendicular to the plane of imaging. When it is oblique it is possible to demonstrate directional flow, but estimation of velocity is more difficult (Bradley and Waluch 1985). The clinical importance of these observations is in determining patent lumina from areas of thrombosis. The information will also be of value in the study of degenerative diseases of the blood vessels where shear effects may affect the vessel walls.

Quantitation of blood flow by NMR is still at an early stage. When it is clinically available it should be a powerful tool without the need for the acoustic windows required by ultrasound or the radiopharmaceuticals required by radionuclide techniques. It will be some time, however, before NMR techniques will compete with radionuclide methods for organ "nutritional" blood flow assessment, especially in view of the recent development of HM-PAO.

Tissue Metabolism

Radionuclide Methods

Radionuclides provide the ideal tracers for following tissue metabolism. Unfortunately, the commonest elements in the body — carbon, nitrogen, oxygen and hydrogen — all have externally detectable radioactive isotopes which have both short half-lives and must be produced by very expensive cyclotrons. This limits their application to rather special research situations where the chemistry is relatively simple and fast (Kairento et al. 1985; Lamerstma et al. 1985; Burns et al. 1984; Patronas et al. 1982; Reiman et al. 1982; Zanzonico et al. 1983). However, increasing numbers of radiopharmaceuticals labelled with longer-lived radionuclides are becoming available for the study of specific aspects of metabolism. The earliest and still the commonest radionuclide is radioactive iodine which is used for the study of thyroid metabolism. The rate of uptake, distribution of function within the organ and the rate of production of thyroid hormones can all be studied either by in vivo or in vitro means. The distribution of inorganic non-radioactive iodine can also be studied by fluorescent techniques (Johnson et al. 1979; Leger et al. 1983) but the results so far have been of limited clinical value. Examples of organs or body systems which can be studied by radioactive techniques are listed in Table 1.1.

Table 1.1. Use of radionuclides in metabolic studies

Organ	Radionuclide	Pharmaceutical	Aspect of metabolism
Brain	0-15	Oxygen	Oxygen
Brain	F-18	Deoxyglucose	Glucose
Thyroid	I-123,131,124	Iodide	T_4 production
Parathyroid	Se-75	Selenomethionine	Amino acid turnover
Heart	Tl-201	Thallium	Myocardial blood flow
Heart	I-131	Fatty acids	Fat turnover
Lung	O-15	Oxygen	Oxygen transfer
Liver	Tc-99m	HIDA	Liver function
Liver	Se-75	SehCAT	Bile salt turnover
Pancreas	Se-75	Selenomethionine	Enzyme synthesis
Kidney	Cr-51,Tc-99m	Chelates	Glomerular filtration
Kidney	I-131,I-123	Hippuran	Tubular function
Skeleton	Ca-45	Calcium chloride	Bone metabolism
Phaeochromocytoma	I-131,I-123	MIBG	Adrenalin synthesis
Bone marrow	Fe-52	Ferric chloride	Red cell production

A more detailed analysis of the movement of the tracers within the body can be made using a compartmental analysis technique based on the clearance or the accumulation measured at organ level. Unfortunately, this type of analysis often does not correlate accurately with a known physiological process. Nevertheless it is obvious that radionuclides provide a powerful and relatively hazard-free method of studying body metabolism in a semi-quantitative or quantitative way.

X-ray and Ultrasound Techniques

While neither X-ray nor ultrasound techniques so far provide similar information on tissue metabolism it is hoped that NMR will be a powerful tool for the study of several aspects of human physiology.

Magnetic Resonance Techniques

The basic information provided by NMR is the distribution of protons, mainly in water and fat, and the degree of binding of the water molecules. The difference between the various normal and abnormal tissues can be seen in the anatomical sections. The relation of the measured parameters to actual tissue metabolism is more obscure.

Areas where NMR has spectacular success include the study of the fat in the brain and abnormalities in multiple sclerosis (Young et al. 1981). The presence of iron results in decreased T_1 and T_2 as the result of its paramagnetic effect. In metabolic diseases of the liver such as haemochromatosis it is possible to demonstrate the presence of the iron by the reduction in relaxation time (Runge et al. 1983). NMR is not appropriate for skeletal metabolic studies due to lack of signal from calcium. It would seem therefore that radioactive techniques are better for measuring mineral turnover in bone.

The technique of NMR spectroscopy, where it is possible to investigate the binding of elements within molecules in vivo, is described in detail in Chap. 4. The technique has been used for many years, both in the laboratory in animals and in human limbs, to investigate many aspects of metabolism. Although it is possible to study several elements in vivo (Feinberg et al. 1985) most NMR work has concentrated on ^{31}P, the naturally occurring isotope of phosphorous. This nucleus gives signals one sixth the strength of the proton. The low natural abundance makes it difficult to make images but it is possible to make spectroscopy measurements in relatively short times due to good signal to noise ratios. The main area of research so far has been into normal and abnormal muscle. In the spectra from normal muscle it is possible to distinguish variations in the level of phosphocreatine and several peaks due to adenosine triphosphate and adenosine diphosphate. The effect on muscle of ischaemia and reoxygenation has been studied, together with the abnormalities found in inherited muscular dystrophy. Studies are now also commencing on the use of spectroscopy for the study of normal and abnormal myocardial muscle.

Other studies have concentrated on neonatal cerebral metabolism, elucidating the effect of ischaemia and oxygenation. It has also been possible to study the effect of hydrocephalus and inherited metabolic defects (Newman 1984). With the new equipment now becoming available which offers the potential of using spectroscopy deep in the body, it should be possible to study metabolism in the liver and in transplanted kidneys. However, at present most effort is concentrating on the use of surface coils and lesions which are amenable to measurement by them. A few studies have taken place on the metabolism of tumors and the effect of chemotherapy on them (Nidecker et al. 1985). The effect of oxygenation on tumours is particularly important because their sensitivity to radiation therapy depends upon their level of oxygenation.

Advantages of NMR

The clinical applications of spectroscopy are considered in detail in Chap. 4. However, in relation to other techniques for assessing tissue metabolism the advantages of spectroscopy are its non-invasiveness and the ability to carry out repeated measurements without hazard, especially in children. No doubt radioactive techniques will be complementary to NMR — each providing unique solutions to particular problems. The difficulties in applying spectroscopic techniques to the clinical situation centre around defining accurately the tissue being measured, especially when using surface coils. When the area of interest contains predominately one tissue, the spectra reflect changes in that tissue. However, in oncology and other clinical situations there are likely to be problems in monitoring responses to therapy due to changes in the composition of the tissues in the area of interest and alterations in the size of the tumour or other pathology before and during treatment.

Tissue Characterisation

Attempts to differentiate tissues on morphological grounds have always proved difficult. This is not surprising when often the diagnosis becomes obvious only when tissues are fixed, sectioned and examined under a microscope. Many lesions can provide a similar morphological appearance on microscopic examination.

Radionuclide Techniques

So far the most successful method of specifying a particular tissue has been with the use of radionuclides specific to that tissue. Radioiodine uptake is the classic example in identifying not only normal thyroid tissue but also well differentiated thyroid cancer. Recently the radiopharmaceutical MIBG has been used to identify chromaffin granule containing tissue including phaeochromocytoma, neuroblastoma etc. (Nakajo et al. 1983).

Antibody imaging has also been used to define specific antigens. When these antigens occur in unexpected areas of the body they can be used to unequivocally identify, for example, the spread of carcinoma of the breast (Rainsbury et al. 1983). At the present time cross-reactions limit the value of this technique but it must be remembered that antibody imaging is still at an early stage of development (Larson 1985; Keenan et al. 1985).

Ultrasound Techniques

Ultrasound has been investigated in depth for its ability to characterise tissues (Thijssen and Nicholas 1982) and two approaches have been employed to date. The first has concentrated on an analysis of the echo pattern seen in typical ultrasound tomographic images. The size, number and spatial distribution of the echoes is the basis of the ability to distinguish between tissues and normal and abnormal areas. While the trained eye is excellent at distinguishing such abnormalities in correctly adjusted images, the hope is that the computer can do better in distinguishing subtle differences. Experience to date has shown that it is difficult to use tissue characterisation to make a firm diagnosis in focal lesions. However, the technique has been valuable in diagnosing diffuse liver disease where it is more difficult to decide visually whether a uniform pattern of echoes is abnormal or normal.

The second approach is to analyse the basic features of ultrasound, i.e. attenuation through tissue, the velocity and the back scattering cross-section. The indications are that this may be more successful in some situations in making absolute tissue diagnosis. While ultrasound has difficulty in making absolute diagnoses, there are several studies which indicate that it is sensitive in monitoring the progress of a disease during therapy.

NMR Tissue Characterisation

NMR tissue characterisation in principle has several advantages over the techniques mentioned so far. Unlike techniques based upon anatomical parameters, the measurements carried out by NMR are related to the chemistry of the tissues under examination. Essentially three parameters are measured: the density of protons, longitudinal and transverse relaxation times. Altering the pulse sequences can produce different weightings of the relaxation times, and, in theory at least, the fact that three parameters are being used instead of one should increase the chances of this approach being more disease- or tissue-specific than others. However, there are also theoretical disadvantages associated with NMR tissue characterisation. These include the fact that since one is measuring signals from protons and their binding in molecules the technique is more sensitive to fluid, oedema and fat rather than to specific tissues. It is rather unlikely that NMR will give a particular histological diagnosis but more likely that it will give an indication of how much fat or fluid is present in the volume of interest. Thus it is usual to have an overlap in values for T_1 and T_2 for diseases or tissues (Mathur-De Vré 1984); in any case there are technical problems in actually measuring the relaxation times with a high degree of confidence (Chap. 9). However while an absolute diagnosis may be lacking, NMR does have a higher sensitivity and specificity than other techniques.

Clinical situations where NMR has already been of value include abnormalities of iron and copper metabolism. Iron accumulation occurs in primary haemochromatosis, transfusion haemosiderosis, and alcoholic cirrhosis. Increased concentrations of copper are found and have been measured by NMR in Wilson's disease, in the liver (Runge et al. 1983), in the brain (Mills et al. 1984a) and are also found in lymphomas and long-standing biliary obstruction. Measurements of relaxation times in brain tumours have been assessed for their ability to predict histological type: they have proved helpful but not specific enough to be of clinical value (Araki et al. 1984; Mills et al. 1984a; Rinck et al. 1985). NMR has been successful in the brain for determining the state of intracranial haemorrhages and haematomata. The changes in relaxation times have followed the changes in fluidity (Swensen et al. 1985; Gomori et al. 1985; Bradley and Schmidt 1985; Sipponen et al. 1984). As with ultrasound, NMR is useful in determining the presence of abscesses due to the fluid contents (Wall et al. 1985) but for specificity the radiolabelled white cell technique is still best. A fluid-containing lesion in the liver which can cause problems in diagnosis is cavernous haemangioma. The ultrasound technique, while sensitive for fluid containing lesions, can give a false positive diagnosis of secondary cancer — probably due to the high level echoes produced by the vessel walls. NMR has proved to be more sensitive than CT or ultrasound for this lesion but differential diagnosis remains difficult (Glazer et al. 1985a). The study of gall bladder contents in fasting and normal animals has enabled deductions to be made about the composition of the bile fluid (Demas et al. 1985). However, it has not so far been possible to use this type of information to differentiate between acute and chronic cholecystitis (Loflin et al. 1985). As might be expected, NMR has been less sensitive than CT scanning to the presence of the calcium normally associated with chronic pathology (Holland et al. 1985) such as tumours and a-v malformations.

In the thorax a recurring problem is the differentiation of fibrosis from recurrent tumour. CT is relatively insensitive, while gallium-67 scans are only partially successful. NMR shows promise, tumours having a higher T_2 value than fibrotic tissue (Glazer et al. 1985b; Rholl et al. 1985). Both CT and NMR scanning have successfully detected and differentiated benign and malignant lipomatous tumours. Benign lesions have shown relaxation values similar to normal subcutaneous fatty tissue (Dooms et al. 1985). NMR thus has some advantages over other techniques for tissue characterisation but the problems of confidently detecting malignant change remain.

Conclusions

This brief chapter has discussed various non-invasive techniques for investigating aspects of body physiology, metabolism and tissue differentiation. With a few exceptions it is still not clear which methods will predominate in future years. It is hoped that the main limitation of NMR investigation, i.e. the cost, will steadily reduce to the point where the several methods can be compared on their scientific merits rather than on economics.

References

Adisehiah M, Barber RW, Szaz KF (1984) Measurement of regional limb blood flow in normal humans by inhalation of [133]Xe. Eur J Nucl Med 9: 379–381

Ahmad M, Fletscher JW, Pur-Shahriari AA, George EA, Donati RM (1979) Radionuclide venography and lung scanning: concise communication. J Nucl Med 20: 291–293

Amparo EG, Higgins CB, Farmer D, Gamsu G, McNamara M (1984) Gated MRI of cardiac and paracardiac masses: initial experience. AJR 143: 1151–1156

Araki T, Inouye T, Suzuki H, Machida T, Ilio M (1984) Magnetic resonance imaging of brain tumors: Measurement of T1. Radiology 150: 95–98

Bassingthwaite JB, Holloway GA (1976) Estimation of blood flow with radioactive tracers. Semin Nucl Med 6: 141–162

Bradley WG, Schmidt PG (1985) Effect of methemoglobin formation on the MR appearance of subarachnoid hemorrhage. Radiology 156: 99–103

Bradley WG, Waluch V (1985) Blood flow: magnetic resonance imaging. Radiology 154: 443–450

Burns HD, Dannals RF, Langstrom B et al. (1984) (3-N-(^{11}C)Methyl)Spiperone, a ligand binding to dopamine receptors: Radiochemical synthesis and biodistribution studies in mice. J Nucl Med 25: 1222–1227

Bursch JH, Heintzen PH (1985) Parametric imaging in digital radiography. Radiol Clin North Am 23: 321–334

Coltart RS, Wraight P (1985) The value of radionuclide venography in superior vena caval obstruction. Clin Radiol 36: 415–418

Conces DJ, Vix VA, Klatte EC (1985) Gated MR imaging of left atrial myxomas. Radiology 156: 445–447

Demas BE, Hricak H, Moseley M et al. (1985) Gallbladder bile: an experimental study in dogs using MR imaging and proton MR spectroscopy. Radiology 157: 453–455

Dooms GC, Hricak H, Sollitto RA, Higgins CB (1985) Lipomatous tumors and tumors with fatty component: MR imaging potential and comparison of MR and CT results. Radiology 157: 479–483

Ennis TJ, Dowsett DJ (1983) Vascular radionuclide imaging — A clinical atlas. John Wiley, Chichester

Feinberg DA, Crooks LA, Kaufman L et al. (1985) Magnetic resonance imaging performance: a comparison of sodium and hydrogen. Radiology 156: 133–138

Fletcher BD, Jacobstein MD, Nelson AD, Riemenschneider TA, Alfidi RJ (1984) Gated magnetic resonance imaging of congenital cardiac malformations. Radiology 150: 137–140

Gill RW (1985) Measurement of blood flow by ultrasound: Accuracy and sources of error. Ultrasound Med Biol II (No 4): 625–641

Glazer GM, Aisen AM, Francis IR, Gyves JW, Lande I, Adler DD (1985a) Hepatic cavernous hemagioma: magnetic resonance imaging. Radiology 155: 417–420

Glazer HS, Lee JKT, Levitt RG et al. (1985b) Radiation fibrosis differentiation from recurrent tumour by MR imaging: work in progress. Radiology 156: 721–726

Glazer HS, Gutierrez FR, Levitt RG, Lee JKT, Murphy WA (1985c) The thoracic aorta studied by MR imaging. Radiology 157: 149–155

Go RT, O'Donnell JK, Underwood DA et al. (1985) AJR 145: 21–25

Gomori JM, Grossman RI, Goldberg HI, Zimmerman RA, Bilanuik LT (1985) Intracranial hematomas: Imaging by high-field MR. Radiology 157: 87–93

Handler CE, Ardley RG, Maisey MN (1985) Gated thallium tomography — potential for improved accuracy in the detection of coronary artery disease. Br J Radiol 58: 107–110

Herfkens RJ, Higgins CB, Hricak K et al. (1983) Nuclear magnetic resonance imaging of atherosclerotic disease. Radiology 148: 161–166

Holland BA, Kucharcyzk W, Brant-Zawadzki MN, Norman D, Haas DK, Harper PS (1985) MR imaging of calcified intracranial lesions. Radiology 157: 353–356

Holman BL, Hill TC, Magistretti PL (1982) Brain imaging with emission computed tomography and radio-labeled amines. Invest Radiol 17: 206–215

Holman BL, Lee RGL, Hill TC, Lovett RD, Lister-James J (1984) A comparison of two cerebral perfusion tracers, N-isopropyl I-123 p-iodoamphetamine and I-123 HIPDM in the human. J Nucl Med 25: 25–30

Johnson PM, Esser PD, Lister DB (1979) Fluorescent thyroid imaging: Clinical evaluation of an alternative instrument. Radiology 130: 219–222

Kairento AL, Brownell GL, Elmaleh DR, Swartz MR (1985) Comparative measurement of regional blood flow, oxygen and glucose utilisation in soft tissue tumor of rabbit with positron imaging. Br J Radiol 58: 637–643

Keenan AM, Harbert JC, Larson SM (1985) Monoclonal antibodies in nuclear medicine. J Nucl Med 26: 531–537

Lamerstma AA, Wise RJS, Cox TCS, Thomas DGT, Jones T (1985) Measurement of blood flow, oxygen utilisation, oxygen extraction ratio, and fractional blood volume in human brain tumours and surrounding oedematous tissue. Br J Radiol 58: 725–734

Larson SM (1985) Radiolabelled monoclonal anti-tumor antibodies in diagnosis and therapy. J Nucl Med 26: 538–545

Leger AF, Fragu P, Rougier P, Laurent MF, Tubiana M, Savoie JC (1983) Thyroid iodine content measured by X-ray fluorescence in amiodarone-induced thyrotoxicosis: Concise communication. J Nucl Med 24: 582–585

Loflin TG, Simeone JF, Mueller PR et al. (1985) Gallbladder bile in cholecystitis: in vitro MR evaluation. Radiology 157: 457–459

Mathur-De Vré R (1984) Biomedical implications of the relaxation behaviour of water related to NMR imaging. Br J Radiol 57: 955–976

Mills CM, Crooks LE, Kaufman L, Brant-Zawadzki M (1984a) Cerebral abnormalities: use of calculated T1 and T2 magnetic resonance images for diagnosis. Radiology 150: 87–94

Mills CM, Brant-Zawadzki M, Crooks LE et al. (1984b) Nuclear magnetic resonance: principles of blood flow imaging. AJR 142: 165–170

Mullani NA, Goldstein RA, Gould KL et al. (1983) Myocardial perfusion with rubidium-82. 1. Measurement of extraction fraction and flow with external detectors. J Nucl Med 24: 898–906

Nakajo M, Shapiro B, Copp J et al. (1983) The normal and abnormal distribution of the adrenomedullary imaging agent m-[I-131] iodobenzylguanidine (I-131 MIGB) in man: evaluation by scintigraphy. J Nucl Med 24: 672–682

Newman RJ (1984) Clinical applications of nuclear magnetic resonance spectroscopy: a review. J R Soc Med 77: 774–778

Nidecker AC, Muller S, Aue WP et al. (1985) Extremity bone tumours: evaluation by P-31 MR spectroscopy. Radiology 157: 167–174

Nowotnik DP, Canning LR, Cumming SA et al. (1985) Development of a $^{99}Tc^m$ labelled radiopharmaceutical for cerebral blood imaging. Nucl Med Comm 6: 499–506

Patronas NJ, Di Chiro G, Brooks RA et al. (1982) Work in progress: [^{18}F] fluorodeoxyglucose and positron emission tomography in the evaluation of radiation necrosis of the brain. Radiology 144: 885–889

Rainsbury RM, Westwood JH, Coombes RC et al. (1983) Location of metastatic breast carcinoma by a monoclonal antibody chelate labelled with indium-III. Lancet II: 934–938

Rall JE (1956) The role of radioactive iodine in the diagnosis of thyroid disease. Am J Med 20: 719

Rehr RB, Malloy CR, Filipchuk NG, Peshock RM (1985) Left ventricular volumes measured by MR imaging. Radiology 156: 717–719

Reiman RE, Benua RS, Gelbard AS, Allen JC, Vomero JJ, Laughlin JS (1982) Imaging of brain tumors after administration of L-(N-13) glutamate: concise communication. J Nucl Med 23: 682–687

Rholl KS, Levitt RG, Glazer HS (1985) Magnetic resonance imaging of fibrosing mediastinitis. AJR 145: 255–259

Rinck PA, Meindl S, Higer HP, Bieler EU, Pfannenstiel P (1985) Brain tumors: detection and typing by use of CPMG sequences and in vivo T2 measurements. Radiology 157: 103–106

Runge VM, Clanton JA, Smith FW et al. (1983) Nuclear magnetic resonance of iron and copper states. AJR 141: 943–948

Saddekni S, Sos TA, Srur M, Cohn DJ (1985) Contrast administration and techniques of digital subtraction angiography performance in digital radiography. Radiol Clin North Am 23: 275–291

Sipponen JT, Sipponen RE, Sivula A (1984) Chronic subdural hematoma: Demonstration of magnetic resonance. Radiology 150: 79–85

Swenson SJ, Keller PL, Berquist TH, McLeod RA, Stephens DH (1985) Magnetic resonance imaging of hemorrhage. AJR 145: 921–927

Switzer D, Nanda N (1985) Doppler color flow mapping. Ultrasound Med Biol 11: 403–416

Taylor K, Burns P, Woodcock J, Wells N (1985) Blood flow in deep abdominal and pelvic muscles: ultrasonic pulsed Doppler analysis. Radiology 154: 487–493

Thijssen JM, Nicholas D (1982) Ultrasonic tissue characterization. Martinus Nijhoff Publishers for the Commission of the European Communities, The Hague

Underwood SR, Klipstein RH, Firmin DN et al. (1986) Magnetic resonance assessment of the accuracy of radionuclide methods for the quantification of valvular regurgitation and atrial shunting. In: Hofer R, Bergmann H (eds) Radioaktive Isotope in Klinik und Forschung, vol 17. Ergmann, Vienna, pp 299–303

Wall SD, Fisher MR, Amparo EG, Hricak H, Higgins CB (1985) Magnetic resonance imaging in the evaluation of abscesses. AJR 144: 1217–1221

Webb S, Ott RJ, Flower MA, Leach MO, Marsden P, McCready VR (1986) Verification of a technique for the measurement of functioning thyroid volume using positron emission tomography. Med Biol Engineering 23 (Suppl 2): 1397–1398

Wells PNT, Skidmore RV (1985) Doppler developments in the last quinquenium. Ultrasound Med Biol 11: 613–623

Young IR, Hall AS, Pallis CA, Legg NJ, Bydder GM, Steiner RE (1981) Nuclear magnetic resonance imaging of the brain in multiple sclerosis. Lancet II: 1063–1066

Zanzonico PB, Bigler RE, Schmall B (1983) Neuroleptic binding sites: Specific labelling in mice with [18] haloperidol, a potential tracer for positron emission tomography. J Nucl Med 24: 408–416

2. The Physical Basis of NMR Studies Measuring Physiological Function and Metabolism

M. O. Leach

Introduction

This chapter provides an introductory description to the physics of NMR suitable for the non-physicists and is intended to provide a framework for the functional and metabolic studies and techniques described in later chapters. The emphasis has been on providing a coherent description that can be applied to all of the techniques covered by this book with more detailed descriptions of particular aspects of NMR being given in appropriate chapters. In addition, a bibliography of introductory articles and books covering both the physics and the applications of NMR has been included at the end of this chapter.

The chapter commences by describing the behaviour of nuclei in a magnetic field, firstly considering isolated nuclei, and then considering the properties of bulk material, and the relaxation properties of the induced magnetism. NMR imaging techniques are described, as are methods of obtaining functional and metabolic information. Finally instrumental, siting and safety considerations are discussed.

Physical Basis of NMR

The Behaviour of Individual Nuclei

The nucleus of an atom is composed of protons and neutrons both of which can be considered as having a distribution of electric charge. The distribution of charge in the nucleus spins about the centre of the nucleus and this spinning, or circulating charge is therefore equivalent to an electric current. Figure 2.1 shows that an electric current circulating in a coil generates a magnetic field perpendicular to the plane of the coil and in the same way the circulating charge in the nucleus, which in the most simple case will consist of just a proton, also generates a magnetic field or magnetic moment. This is known as the nuclear magnetic moment. For a nucleon heavier than

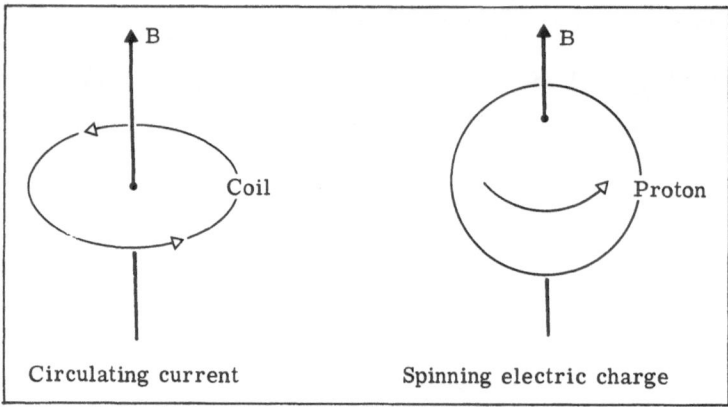

Fig. 2.1.   Circulating current produces a magnetic field.

hydrogen, the magnetic moment is the vector sum of the individual magnetic moments of each of the nucleons composing the nucleus of the atom.

The nucleons of the nucleus can be considered as occupying discrete orbitals or more accurately, energy states, with neutrons and protons filling separate allowed orbits, this being a consequence of the exclusion principle in quantum mechanics. In nuclei with even numbers of protons and neutrons, the nucleons usually occupy the available orbits or states in pairs, and can be considered as rotating in opposite senses as shown in Fig. 2.2. When both orbitals in a pair of orbits are filled, there is no net magnetic moment from those nuclei, so magnetic nuclei are therefore those having an odd number of either protons, neutrons or both, e.g. ^1H, ^{13}C and ^2D. From now on therefore only nuclei having a net magnetic moment and therefore having an odd number of either protons and/or neutrons will be considered.

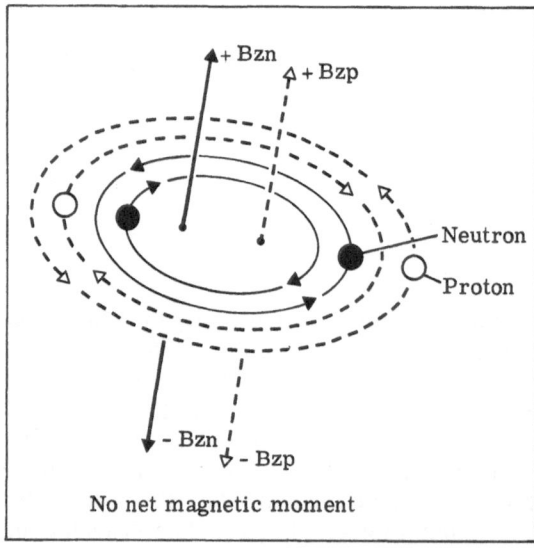

Fig. 2.2.   Paired nuclei in a nucleus.

In the presence of an external magnetic field, the nuclei having a magnetic moment can orientate themselves either with or against the direction of the magnetic field, and so occupy a spin up or spin down state. For each pair of orbits one will be available to a nucleon with spin up and the other to a nucleon with spin down. It is not possible to have two spin up or two spin down nuclei occupying the same pair of orbits.

There is a very small difference in energy between the spin up and the spin down states and a transition from one state to the other can be induced by irradiating the nucleus with very small amounts of energy at the NMR Larmor precession frequency for the nucleus. The resonant frequency

$$\nu_0 = \frac{\gamma}{2\pi} B_0 \qquad (1)$$

where γ is the nuclear gyromagnetic ratio, and B_0 is the magnetic field strength.

Figure 2.3 shows the process of resonant absorption and loss of energy by nuclear magnetic resonance. When energy is absorbed a nucleon having spin up will transfer

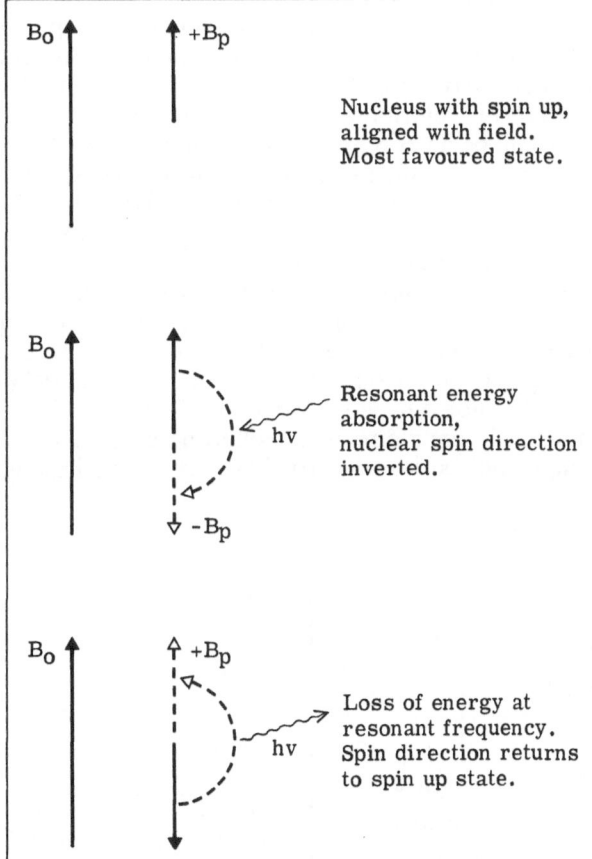

Fig. 2.3. Resonant absorption and loss of energy by a proton.

to the other orbit available to it in its pair of orbitals and will then be in a spin down state. In order for this to occcur some energy must be supplied to the nucleus. Likewise the nucleus can return to the spin up state by losing energy at the resonant frequency.

The magnetic moment of the nucleus has so far been considered as being either aligned with or against the external magnetic field. In fact, in the presence of an external magnetic field B_0, the magnetic moment of the nucleus is not exactly aligned with the z axis, but it is orientated at a constant angle to the z axis, determined by quantum mechanics, and precesses about the z axis at the Larmor precession frequency. There is, therefore, a component of the magnetisation in the z direction M_z, and a component of the magnetisation in the x-y plane M_{xy}.

The Behaviour of Nuclei in Bulk Material

In tissue one is dealing with an assembly of a great many nuclei and therefore only the summed magnetic moment from all of the assembly of nuclei is observed, rather than the individual magnetic moments of particular nuclei. In the absence of an external magnetic field there will be no net magnetism in the sample and the spins will be oriented in random directions as shown in Fig. 2.4. In the presence of a magnetic field with no other stimuli (Fig. 2.5) there will be an equilibrium distribution of nuclei between the two spin states with a very slight population excess in the spin up state, that is aligned with the main magnetic field. At room temperature the difference between the two populations is of the order of 1 in 10^6.

When the contributions of all of the individual nuclei are summed, at equilibrium there will be a net magnetic moment along the +z axis resulting from the summed contributions of the M_z component from all of the individual nuclei. If the sample is then irradiated with a circularly polarised radiofrequency (RF) electromagnetic field at the resonant frequency, energy will be supplied to the nuclei, the individual photons having an energy of about 4×10^{-7} eV.

The net M_z component of magnetism, initially orientated along z, precesses about the z axis at the resonant frequency, with the angle between M, the induced magnetism, and the z axis gradually increasing. Depending on how much RF power is supplied, the net magnetic moment of the material will be rotated, so that following a 90° RF pulse directed along the x axis, there will be no component of the magnetic moment in the z direction. Following a 180° RF pulse, the net magnetism of the sample will be inverted so that it is directed along the −z axis and is of equal

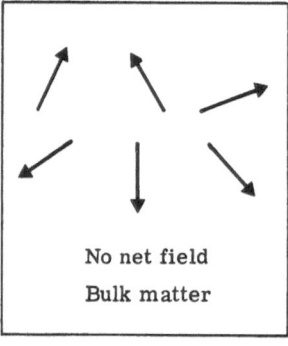

No net field

Bulk matter

Fig. 2.4. Alignment of protons in a sample in the absence of an external magnetic field, showing that no net field from the sample is present.

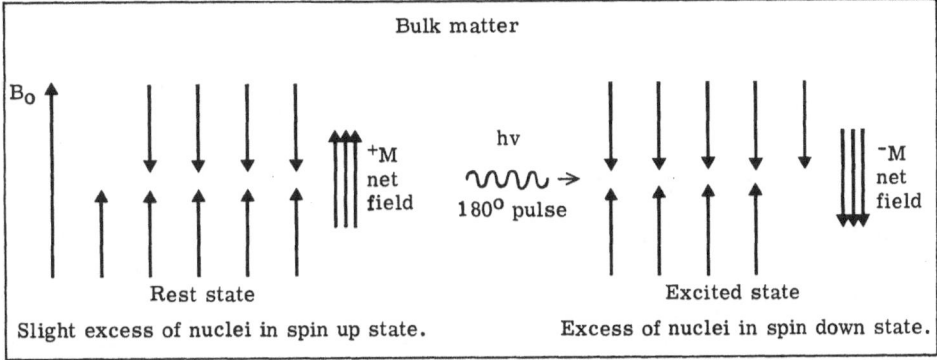

Fig. 2.5. Distribution of protons in spin states in the presence of a B_0 field, before and after a 180° RF pulse.

magnitude to the equilibrium magnetisation. Therefore following a 90° pulse the z component of the magnetisation will be zero and there will be a component in the x-y plane equal to the original equilibrium z magnetisation. An NMR signal can only be detected from magnetisation in the x-y plane and therefore the amount of magnetisation in the z direction must be measured by using 90° pulses to tilt the z component of magnetisation into the x-y plane.

Rotating Frame of Reference

A common method of describing NMR phenomena is to use the rotating frame of reference. To the observer in the laboratory, the net component of magnetism in the x-y plane is rotating around the B_0 direction, the +z axis, at the resonant frequency (42 MHz at 1 Tesla). The convenient way of describing changes in net magnetisation is to consider them in terms of a rotating reference frame which rotates at the Larmor precession frequency. This is equivalent to the view from a position at the end of the net magnetisation vector and very much simplifies descriptions of NMR processes. In the rotating frame of reference all of the magnetisation which is in phase with the exciting RF radiation will appear as a stationary line in the x'-y' plane. Lower frequencies will appear to move anti-clockwise and higher frequencies will appear to move clockwise. The whole frame of reference, i.e. the x'-y' plane, is in fact rotating at the Larmor precession frequency, but the use of the rotating frame description renders this transparent.

Figure 2.6 shows the movement of the magnetisation vector in the rotating frame of reference following an arbitrary α° RF pulse. As can be seen, the z component of the magnetisation is initially equal to the total magnetisation M. However, following the α° pulse, the magnetisation moves to an angle α with respect to the z axis, producing a reduced z component of magnetisation. An x' component of magnetisation can now also be observed in the x-y plane. Figure 2.7 shows the same process but described in the laboratory frame of reference. The magnetisation vector is therefore shown as precessing around the z axis. Following an α° pulse, the angle between the magnetisation vector and the z axis increases as it traces a path along the surface of the sphere. Following a 90° pulse the magnetisation rotates until it lies in the x-y

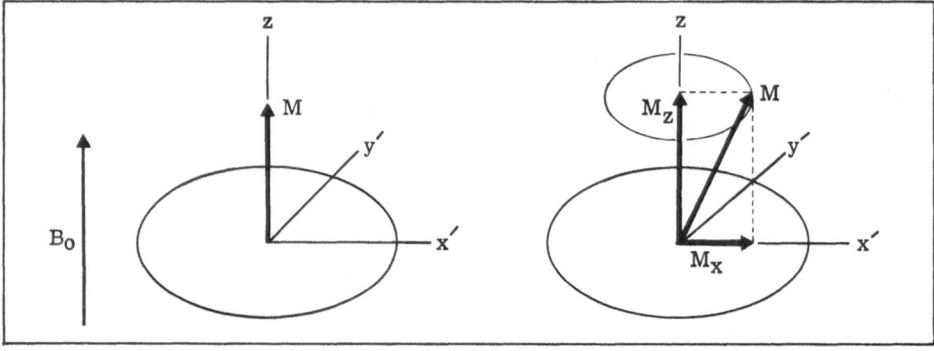

Fig. 2.6.   The change in the direction of net sample magnetisation M following an α° RF pulse, shown in the rotating frame of reference.

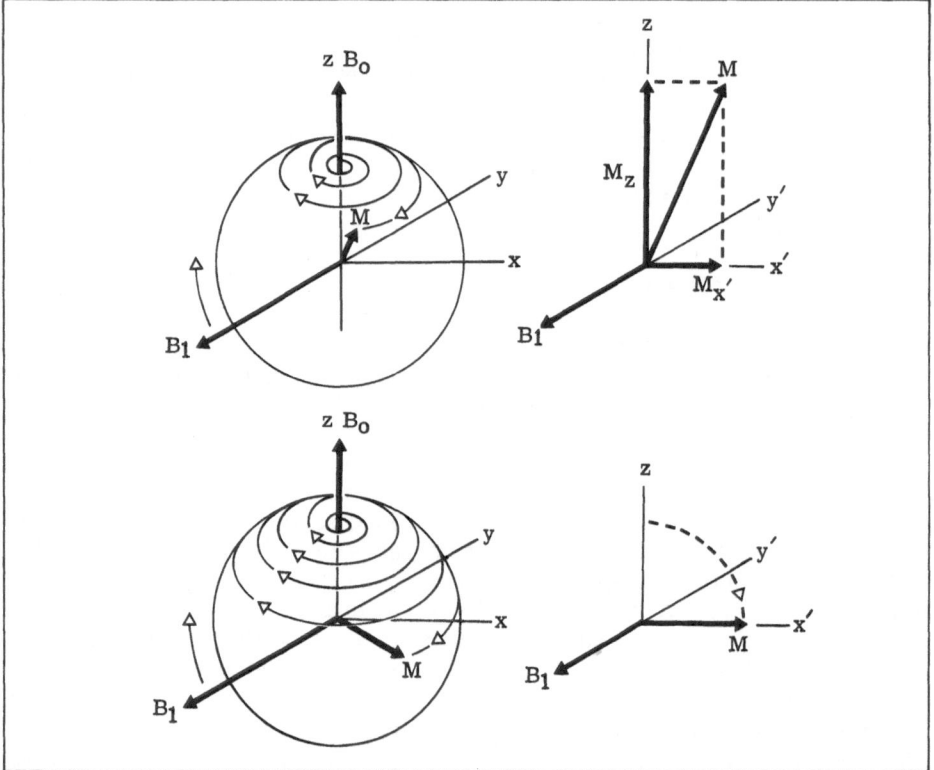

Fig. 2.7.   The result of applying α° RF pulses shown in the laboratory and rotating reference frames.

plane. The equivalent diagrams for the rotating frame of reference are also shown. Figure 2.8 shows the disturbance of the equilibrium magnetisation following both a 90° pulse and a 180° pulse, and also shows schematically the changes in the relative spin populations in the spin up and spin down states.

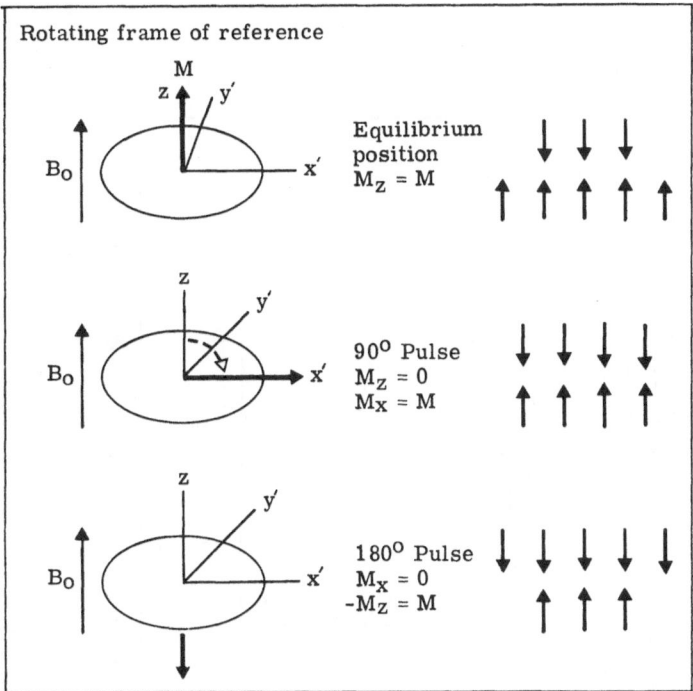

Fig. 2.8. The change in equilibrium magnetisation following 90° and 180° RF pulses. The change in spin populations is shown schematically.

T_1 or Longitudinal Relaxation

Once the net magnetisation has been rotated into the x-y plane by a 90° pulse, the z component of the magnetisation gradually recovers, due to a gradual loss of energy from the nuclei to the surrounding matrix or lattice. This exchange occurs by emission or transfer of energy, causing the spins to revert to their equilibrium state, and is known as T_1 relaxation and is characterised by the T_1 or longitudinal relaxation time. This process is shown in Fig. 2.9. The loss of magnetisation results in a decrease in the angle α between the direction of the net magnetic moment and the z axis. This process is also accompanied by a more rapid dephasing of the x-y component of magnetisation, due to magnetic field inhomogeneities and T_2 relaxation processes (described below). The residual magnetisation in the x-y plane may be large but will not generate a signal if it is de-phased, as the individual magnetic moments of the nuclei will cancel one another out. However the de-phased magnetisation will still produce a net z component. Figure 2.9 also shows the gradual recovery of the z component of the magnetism with time following a 90° RF pulse. Figure 2.10 shows how the recovery of the z component of the magnetisation following an initially 90° pulse can be sampled by using a second 90° pulse at different times (τ) after the initial pulse. The 90° pulses tip the magnetisation into the x-y plane and thus generate a free induction decay which can be measured. The initial amplitude $A(\tau)$ of this is proportional to M_z. Any residual x-y component from

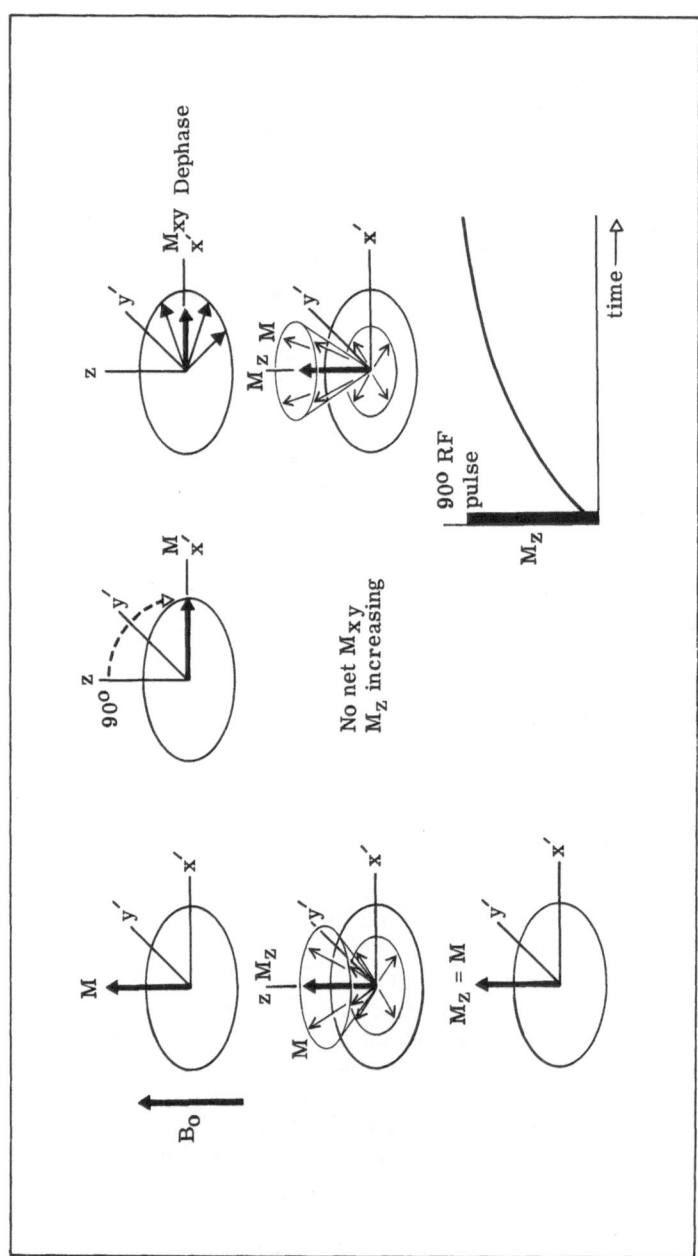

Fig. 2.9.   Longitudinal and transverse relaxation following a 90° RF pulse. The recovery of the M_z component with time is also shown.

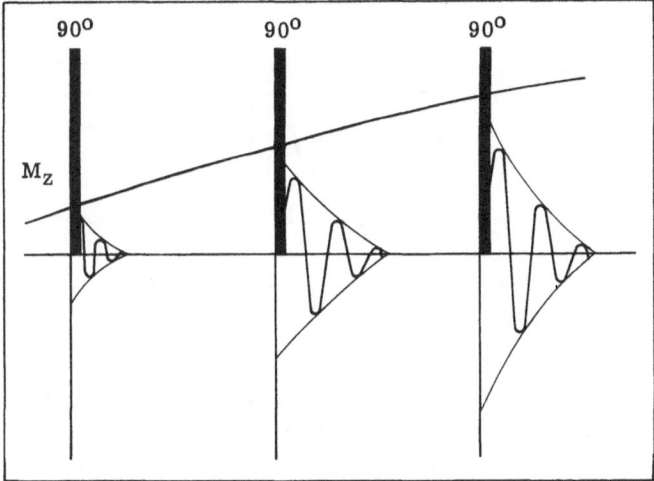

Fig. 2.10. Measuring the M_z component at different times after an initial 90° RF pulse.

previous 90° pulses will not be evident as it will have de-phased by this time. However, if M_z is sampled before complete T_1 relaxation has occurred, M_{xy} will be smaller than the equilibrium value $A(0)$. This process of sampling M_z allows T_1 to be measured using the relationship

$$A(\tau) = A(0) \cdot [1-\exp(-\tau / T_1)] \qquad (2)$$

T_2 or Transverse Relaxation

Following a 90° pulse directed along the y' axis, the magnetisation is tipped into the x-y plane. Initially all of the nuclei are in phase and their magnetic moment can be described by one vector along the x' axis. If all the nuclei precessed at exactly the same speed, that single vector would continue to describe them all. However, in a real sample, all of the nuclei experience slightly different magnetic fields due to molecular magnetic fields and to inhomogeneities in the main B_0 field. Thus they will all precess with slightly different frequencies. This results in de-phasing, with the signals spreading out in the x-y plane, as shown in Fig. 2.11 where the increasing spread of the magnetic moments of the nuclei in the x-y plane results in a reduction in the x-y component M_{xy} of the sample magnetisation. This signal in the x-y plane, the free induction decay, can be detected with a suitable coil and is shown in Fig. 2.11. The exponential fall-off of the free induction decay is due to both T_2 relaxation and magnetic field inhomogeneities, and is known as the effective T_2 or T_2^*.

The transverse relaxation can be measured in the following way. After a 90° RF pulse, spins in the x-y plane de-phase due to local field inhomogeneities and to T_2 relaxation processes as shown in Fig. 2.12. If after a time τ a 180° RF pulse is applied along the y' axis the spins rotate through 180° about the y' axis and are then still in the

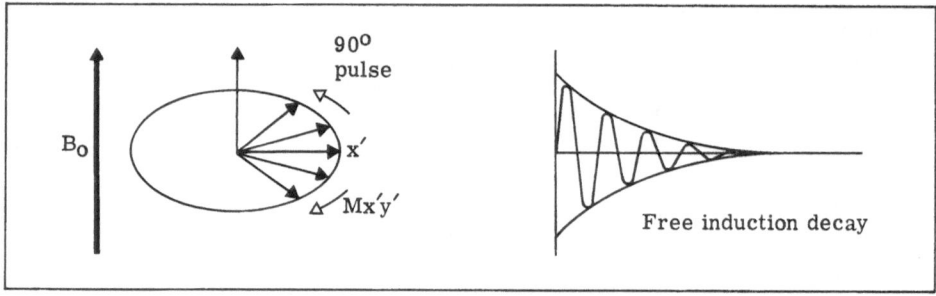

Fig. 2.11.   Transverse relaxation and the free induction decay produced after a 90° RF pulse.

x-y plane but converging or re-phasing rather than de-phasing. In re-phasing they produce an echo signal $A(2\tau)$ at time 2τ. During the re-phasing period, nuclei will be subject to the same magnetic field inhomogeneities as they were during the de-phasing period and so those precessing rapidly will continue to do so, as will those precessing at lower speeds. Thus re-phasing will cancel out the phase shifts due to inhomogeneities in the B_0 field. A useful analogy is to consider the nuclei as a group of runners (this is also shown in Fig. 2.12) where if the runners run for a fixed time period t_1, a slow runner will move a short distance and the fast runner a long distance in that time period. If after t_1 they then reverse direction and run at the same speeds for another period t_1, they will both arrive back at the origin at the same time and therefore will again be in phase. Changes in frequency due to molecular diffusion will not be cancelled out by this 180° pulse process and will result in an overall loss of phase and a reduction in net x-y magnetisation. 180° pulses can be repeated to give multiple echoes but with a decay in the maxima of the echoes (see Fig. 2.12) with time which is described by the T_2 relaxation time. The much faster decay of the individual echoes is described by the T_2^*, or effective T_2 relaxation time, which also includes the effects of magnetic field inhomogeneities.

The echo amplitude at time 2τ is given by

$$A(2\tau) = A(O) \exp \left[-\frac{2\tau}{T_2} - \frac{2}{3}\gamma^2 G^2 D \tau^3 \right] \tag{3}$$

where G is the spatial magnetic field gradient and D is the diffusion coefficient. If diffusion effects are significant, errors in measuring T_2 will occur, particularly for large values of τ. This problem can be reduced by using the Carr-Purcell sequence (Carr and Purcell 1954) where a sequence of 180° pulses are used, giving multiple echoes of alternate phase. The accuracy is further improved by using the Carr-Purcell Mieboon-Gill sequence (CPMG) which corrects for non-uniformities in the irradiating B_1 field (Meiboon and Gill 1958). In this modification, the 180° pulses are applied along the y' axis, $\pi/2$ out of phase with the initial 90° pulse, rather than along the x' axis.

Spectroscopy

A nucleus bound to a molecule is subjected to an average magnetic field resulting from the charge distribution in the molecule, which depends upon the particular molecular compound and upon the location of the nucleus within the molecule. This

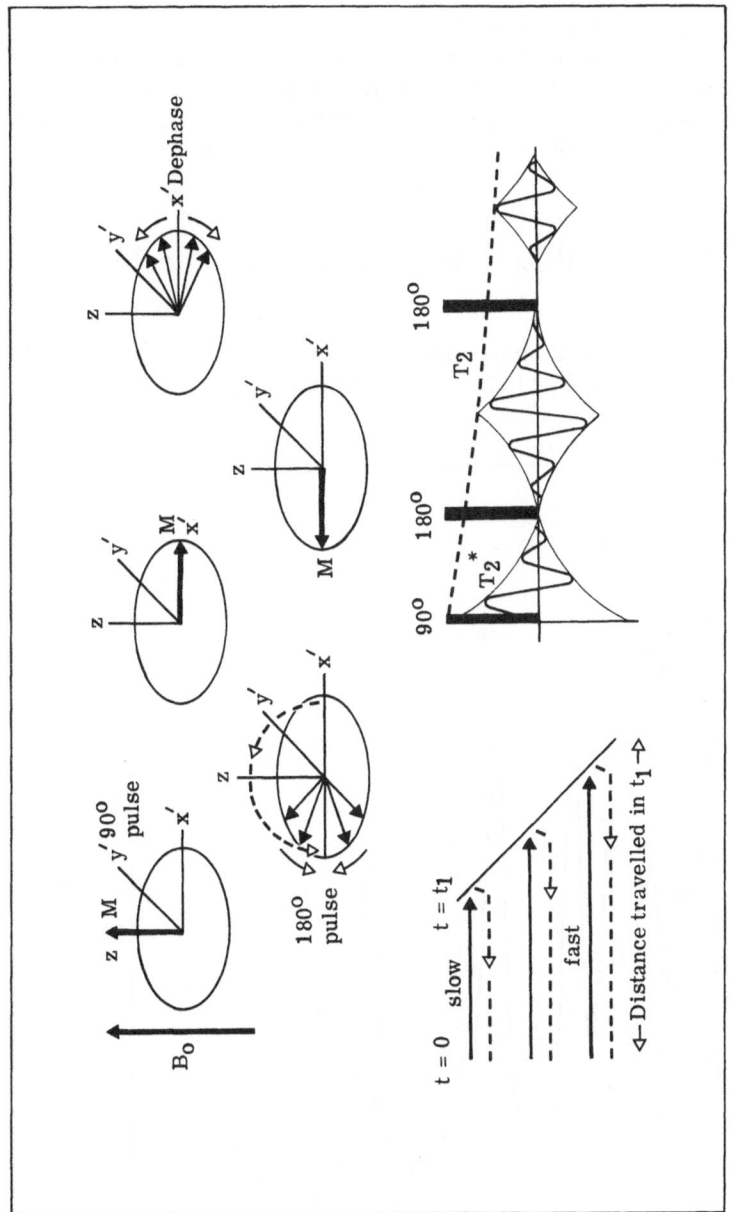

Fig. 2.12. The production of echoes following 180° RF pulses, due to rephasing of the x-y plane magnetism M_{xy}. The analogy of a group of runners and the decay of the echoes is also shown.

causes a slight characteristic shift in resonant frequency. Provided that other effects causing frequency shifts are sufficiently small and that the shift is large enough, the molecular species and molecular position can be identified and presented as a spectrum. In vivo NMR spectroscopy is most usually concerned with measuring ^{31}P, but measurements of ^{1}H, ^{13}C and ^{19}F have also been reported. A typical ^{31}P spectrum is shown in Fig. 2.13. The separation of spectral lines depends on the main B_0 field, and the width of each peak depends on $1/T_2$ and also on the homogeneity of the B_0 field. High homogeneities of 0.1 ppm or greater are desirable, together with magnetic field strengths of at least 1.5 Tesla. Signals are normally collected from a relatively large volume and obtaining a signal from a well localised region is less straightforward.

Imaging and Spatial Localisation of NMR Spectra

To obtain spatially localised image or spectral information, the origin of the NMR signal has to be localised. The relationship between magnetic field strength and fre-

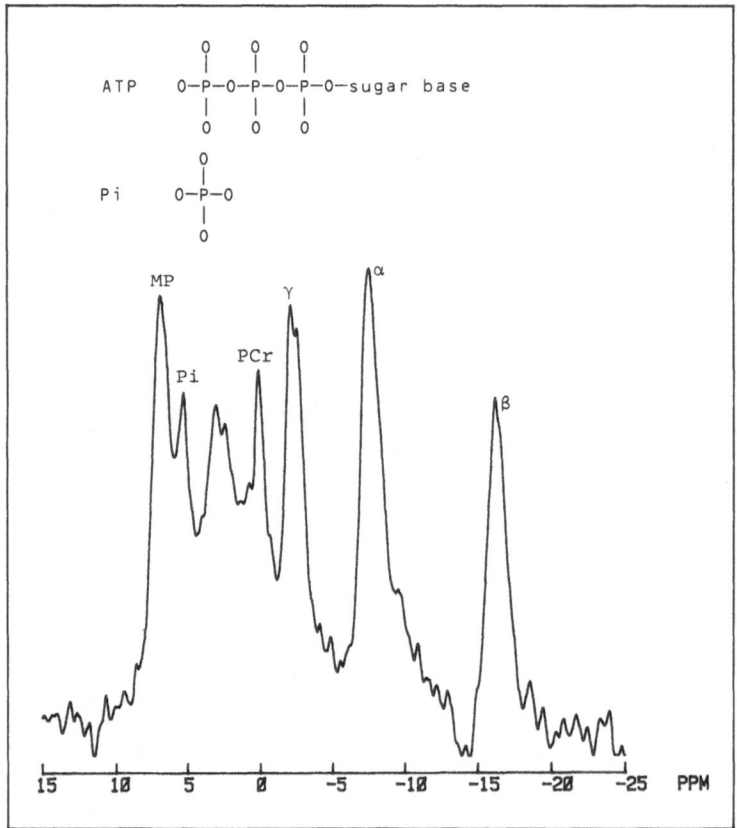

Fig. 2.13.   A typical ^{31}P spectrum, together with the molecules giving rise to the ATP α, ß and γ peaks, and the inorganic phosphate peak.

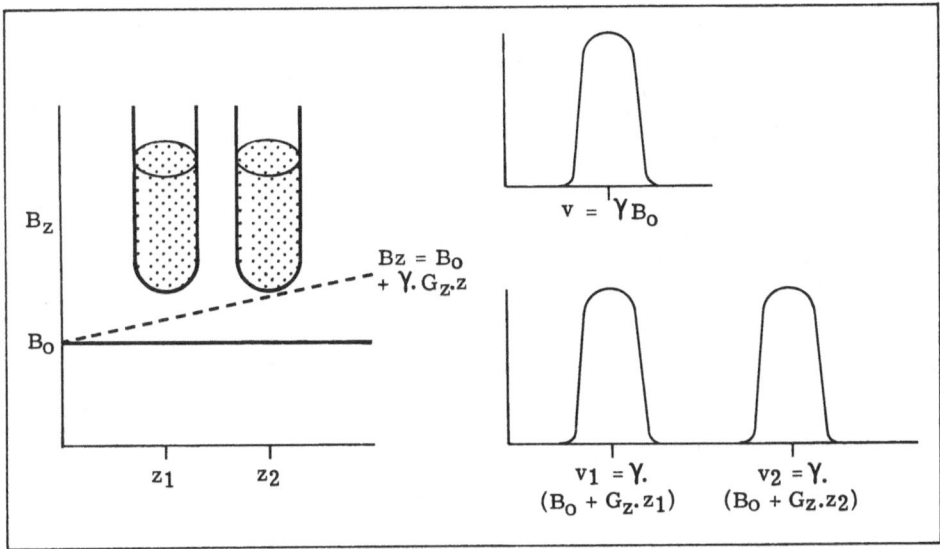

Fig. 2.14.   The effect of a field gradient in modifying the resonant frequency as a function of distance.

quency can be exploited by using additional magnetic field gradients superimposed on the main field. The use of a linear field gradient, which produces a frequency change that is proportional to distance, is the simplest method of producing spatial localisation.

Slice Selection

If a z gradient field is applied and the sample is irradiated with a 90° pulse in a narrow frequency band, only those nuclei experiencing a magnetic field giving a resonant frequency within the irradiating frequency range will be affected by the RF pulse. Figure 2.14 shows the effect of a linear gradient. If no z gradient is present, two test tubes of water at different z positions are subject to the same magnetic field and therefore have the same resonant frequency. If the frequency of the received signal is measured, only one frequency will be evident. If a linear z gradient B_z is superimposed on B_0 the resonant frequency will vary linearly with z displacement, and the two test tubes will produce signals having different frequencies, with the frequency difference being proportional to their separation. Thus if a sample were irradiated in the same way, a slice perpendicular to the z axis with a slice width proportional to the irradiating frequency band width as shown in Fig. 2.15 would be excited, with the net magnetisation in the slice being tipped into the x-y plane to produce a free induction decay (FID).

Projection Reconstruction

If the z gradient is now turned off and the x gradient is turned on, the frequency of the FID will depend upon the x displacement. The FID is recorded as a function of time

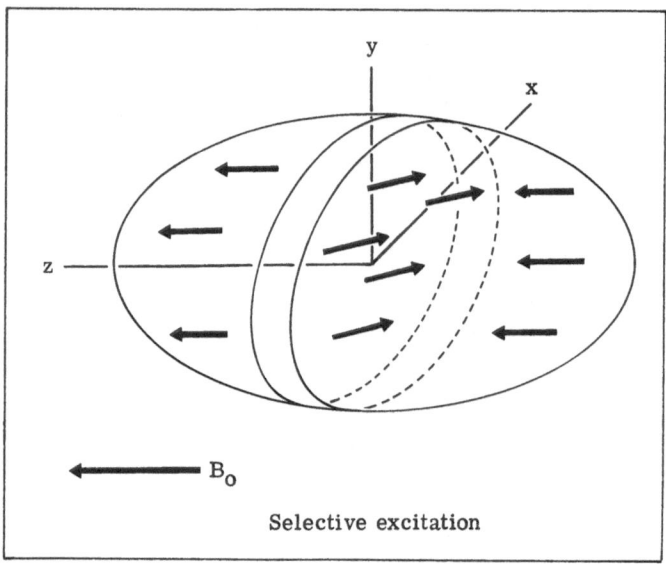

Fig. 2.15. The use of a z field gradient to selectively excite a slice in a sample.

(signal amplitude v time) and can be transformed to appear as a function of frequency by performing a Fourier transform. As frequency is proportional to x displacement, these frequency spectra are then projections of proton density values in the y direction (a summation along the y direction) projected onto the x axis. This is shown in Fig. 2.16. To obtain an image from these projections the data must be treated in the same way as conventional CT projection data (Hounsfield 1973) and an image may be reconstructed by convolution and back projection. This method is often termed zeumatography (Lauterbur 1973). Figure 2.17 show a pulse sequence that can be used to obtain projection reconstruction data. A 90° pulse combined with a z gradient is used to select a slice. This is followed by signal readout in the presence of an appropriate combination of x and y gradients, and these can be varied to produce a set of projections. Different pulse sequences can be used by expanding this sequence to include 180° pulses, with measurement of the echo in the presence of a readout gradient providing projection information. Multislice data can be obtained by interleaving the excitation and measurement of different slices. The projection reconstruction method is now not commonly used on commercial equipment, as the 2D Fourier transform method has been found to give superior results.

2D Fourier Transform Imaging

In 2D Fourier transform imaging, sometimes known as Fourier zeumatography (Kumar et al. 1975a, b), a slice is selected as in projection imaging by using a selective excitation. Frequency encoding is again used but in this case along a fixed axis (x) by applying a readout magnetic gradient. The signal obtained represents a projection onto the x axis. The information in the y direction is phase encoded by applying a y gradient for a known period of time. The frequency and phase encoding directions are shown in Fig. 2.18. Spins subjected to a field gradient precess at different frequencies depending on their position. Thus in a given time spins at different y positions will have rotated by different angles θ_y relative to those at $y = 0$. The difference in angle (or phase) can be measured and for a linear gradient it is proportional to the y displacement. The phase information is obtained by cycling the y gradient through as many values as there are sample points or pixels in the y direction. The image is then obtained by Fourier transforming the FIDs to give the x displacement and then Fourier transforming across the FIDs (phase information) to give the y information. Figure 2.18 also shows a typical imaging sequence for 2D Fourier transform imaging, in this case using a spin echo sequence. The phase encoding gradient Gy can be seen and in this sequence the 180° pulses are normally alternated between positive and negative 180° pulses.

A related technique has been proposed by Hoult (1979) in which the use of phase encoding gradient fields is replaced by the use of a gradient in the irradiating B_1 RF field, resulting in a change in flip angle through the sample.

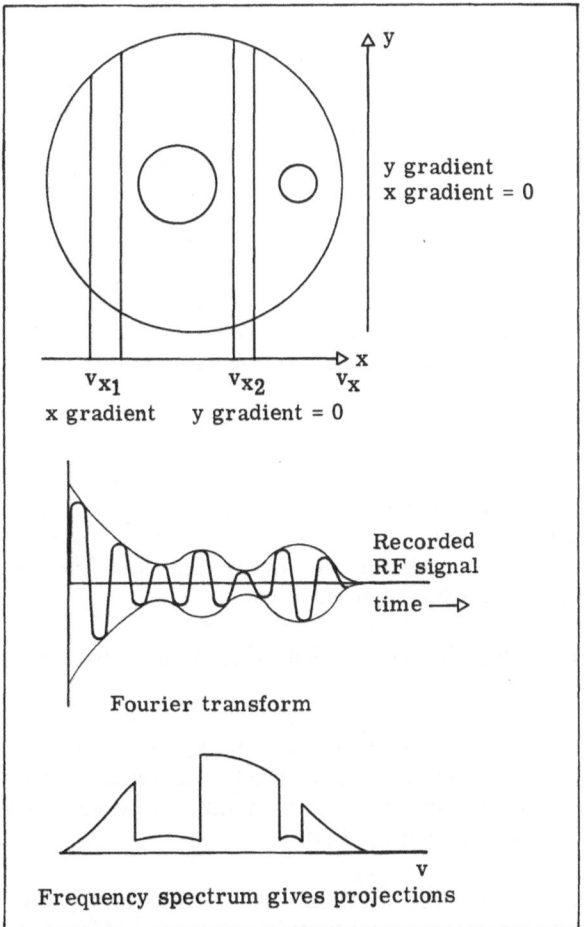

Fig. 2.16. The use of x, y gradient to provide a projection of density information across an object, following Fourier transformation of the free induction decay.

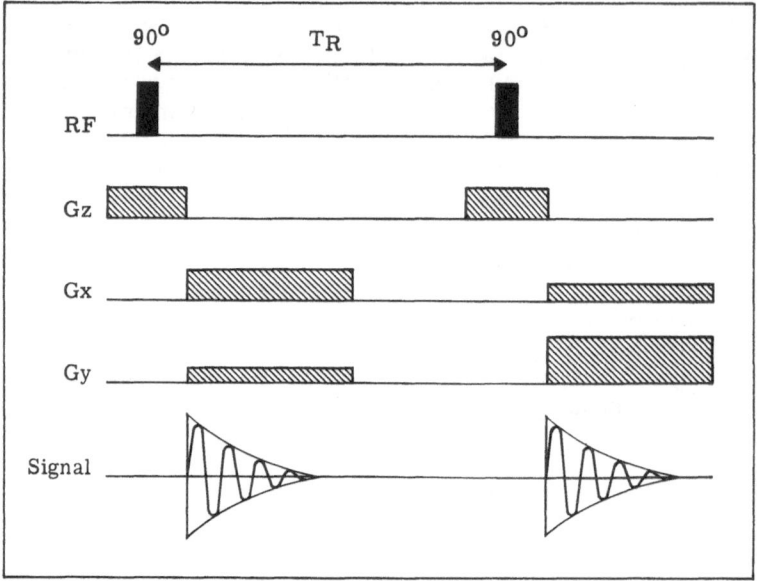

Fig. 2.17. A simple saturation recovery sequence for projection reconstruction imaging.

Other Imaging Methods

3D Fourier Transform. In this method y and z gradients are stepped to provide phase information in y and z with readout using an x frequency encoding gradient (Lai et al. 1981). Information is obtained from the whole sample, but measurement times are long compared with multislice imaging.

Echo Planar Imaging. Following selective excitation with the z gradient, a y gradient is imposed providing frequency encoding and the x gradient is rapidly alternated to produce a series of stimulated echoes giving information at different positions (Mansfield 1977). This is essentially a line scanning, rather than a plane scanning, technique. The method permits much more rapid measurements than is possible with the 2D FT method, although with some reduction in sensitivity.

Spin Warp Imaging (Edelstein et al. 1980). This is a variant of 2D Fourier transform imaging.

Point Scanning (Hinshaw 1976; Damadian et al. 1976) *and Line Scanning* (Andrew et al. 1977; Mansfield et al. 1978). These techniques can be used to obtain signal from localised regions and can be used to build up planar information. However, they suffer from low-sensitivity compared with planar methods.

Comparisons of different image acquisition methods (King and Moran 1984; Mansfield and Morris 1982) and their relative signal to noise properties (Mansfield 1981; Brunner and Ernst 1979) have been reported. Image quality can be considered in two parts. Firstly, the image signal-to-noise, which can be experimentally

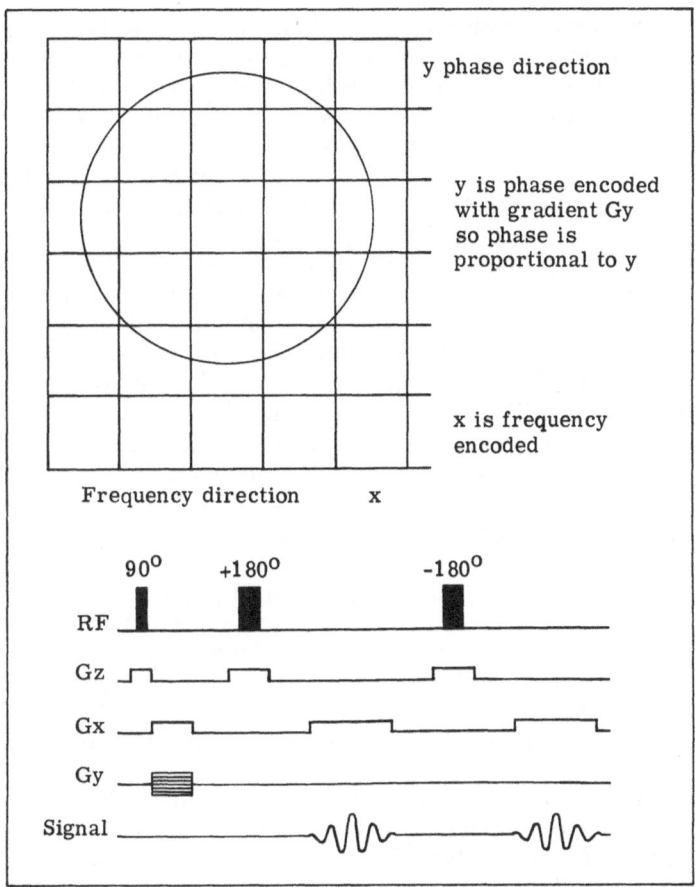

Fig. 2.18. The arrangement of frequency and phase encoding gradients commonly used in 2D Fourier transform imaging, and a simple spin echo sequence suitable for 2D FT imaging.

measured in terms of image noise. This depends on numerous instrumental factors and object properties including relaxation times, B_0 field strength, gradient field strength, pulse sequence, slice thickness and profile, pixel size and signal filtration. Consequently it should be used with caution as an index of instrument performance. Secondly, image degradation due to artefacts may result from patient movement, external RF interference, or various instrumental factors, and the effect of these factors often outweighs theoretical predictions of image noise. Thus, although theoretical predictions suggest that convolution and backprojection reconstruction gives the better signal-to-noise, in fact 2D Fourier transform reconstruction has proved most acceptable in producing high quality images.

Spatially Resolved NMR Spectroscopy (chemical shift imaging)

In in vivo NMR spectroscopy, a spectrum is acquired from a localised region. This region may be defined by a set of field profiling coils which destroy the homogeneity of the B_0 field in all but an approximately spherical volume, a technique known as topical magnetic resonance. Alternatively, a surface coil can be used. This exhibits a rapid fall off in sensitivity, so that signal from a hemispherical-shaped region close to the coil predominates. This approach has been remarkably successful in obtaining signals from relatively superficial locations and may be combined with topical magnetic resonance where greater selectivity is required. Further localisation can be obtained by using depth and refocussing pulse sequences (Bendall and Gordon 1983).

These approaches suffer from the limitation that the sensitive area is not well defined and that signal is only obtained from one region, when it would be an advantage to obtain signal from a number of adjacent areas. In some instances, particularly in oncology, it is desirable to localise the region of interest and one approach is to use a dual-tuned coil (Leach et al. 1985) to provide a proton image prior to carrying out spectroscopy with the same coil remaining in situ.

A variety of spatially selective spectroscopy techniques have recently been developed which can be used in combination with suitable coils. These include depth resolved surface coil spectroscopy (DRESS) (Bottomley et al. 1984), where a selective gradient excitation is used to define a disc-shaped region in the sample. This has been recently extended to provide signals from multiple, localised levels (Bottomley et al. 1985). Another method, which employs an RF gradient technique, is rotating frame spectroscopy (Cox and Styles 1980). An extension of 3DFT imaging to provide localised spectroscopy has been reported by Maudsley et al. (1983). Ordridge (1985) has recently reported an alternative method, image selected in vivo spectroscopy (ISIS), where a combination of selective pulse sequences yield signals which, when appropriately combined, cancel signals from any region outside of a pre-selected cube.

To date, these methods have principally been applied to phosphorus spectroscopy. Several methods have also been proposed for separating water and lipid images in proton imaging (Dixon 1983; Haase et al. 1985). These methods, as well as providing separate water and fat images, also permit the chemical shift artefact due to the different resonant frequencies of water and fat to be removed from the combined image.

Functional Parameters

Relaxation

T_1, Longitudinal or Spin-Lattice Relaxation

T_1 relaxation results from a net exchange of energy with the lattice. This requires a transfer of energy at the resonant frequency from the protons to the lattice and involves spin state changes from spin down to spin up. Molecules, for example water, have a small magnetic moment resulting from the distribution of electric charge within the molecule. The x-y component of this magnetic moment changes as the molecules vibrate, as shown in Fig. 2.19, and if the frequency of the x-y component overlaps with the resonant frequency, energy exchange can occur. This relaxation results in a recovery of the z component of magnetisation following a 90° or 180° pulse.

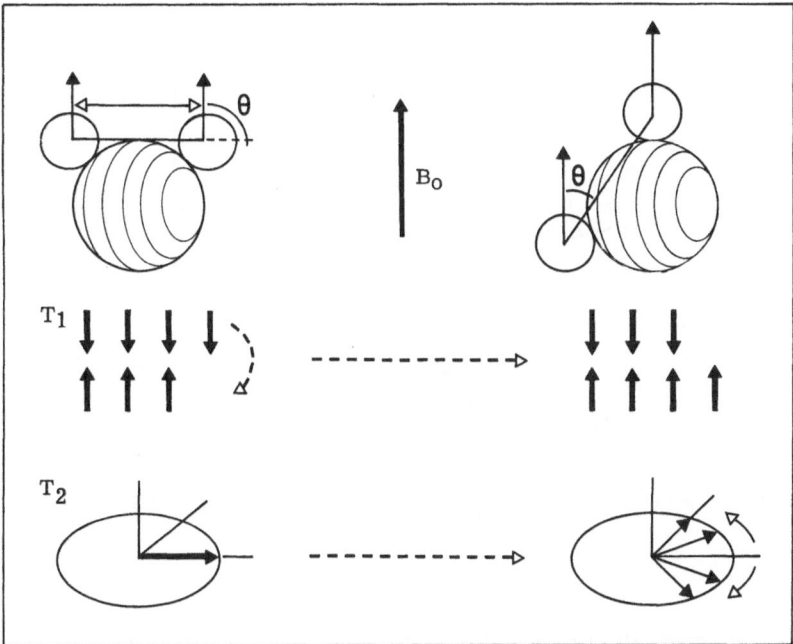

Fig. 2.19. The change in orientation of water molecules relative to the B_0 field changes the z and x-y components of their magnetic field. Changes in the z component result in phase changes, leading to T_2 relaxation; changes in the x-y component result in spin state transitions, producing T_1 relaxation.

T_2, Transverse or Spin-Spin Relaxation

The z component of molecular magnetic fields will cause changes in the precession frequency of protons as a result of modifying the local magnetic field. This causes de-phasing and thus loss of signal in the x-y plane which cannot be recovered with a $180°$ spin echo pulse. T_2 relaxation is the overall loss of signal in the x-y plane.

Blood Flow

NMR images intrinsically exhibit a high degree of blood flow information providing good vessel contrast. For instance, if a selective $90°$ slice excitation is performed, the z component of magnetisation in the plane is set to zero. However, blood flowing into the plane will still have an equilibrium z component and will produce a large signal if interrogated with a $90°$ pulse, giving a high signal compared to the other tissue in the slice, thus providing high vessel contrast. In spin echo sequences, blood flowing into the plane will have no x-y component and will thus produce no echo, again giving good contrast. In multislice imaging and with multiple echoes the process is more complex and both high and low signals can result (Bradley and Waluch 1985; Mills et al. 1983).

These blood flow related phenomena allow special tagging flow sequences to be used, where the flow of an excited bolus of blood through several planes is followed

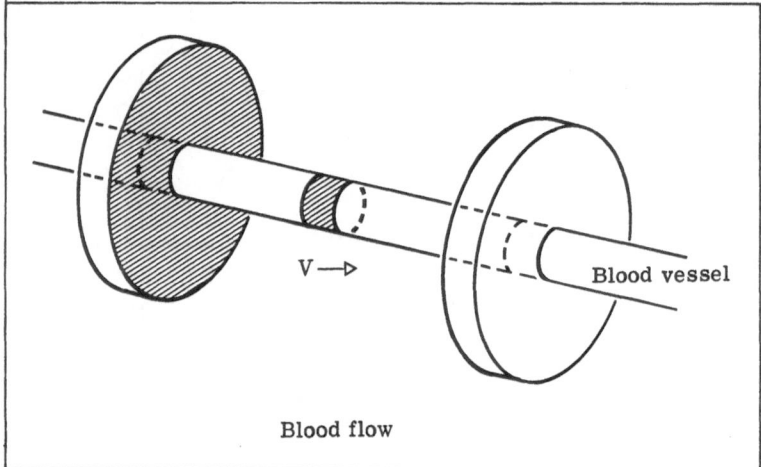

Fig. 2.20. Blood flow in major vessels may be investigated by selective excitation and readout of signals from different planes.

by multi echo sequences, allowing the time of flight between planes of the blood in vessels with a large blood flow to be measured (Fig. 2.20) (Singer 1978; Crooks and Kaufman 1984). These methods give information on major vessel blood flow and may be used in conjunction with cardiac gating to monitor flow changes through the cardiac cycle (George et al. 1984).

Measurement methods based on the changes in phase information that occur as a result of flow are being developed (O'Donnell 1985; Moran et al. 1985). Perfusion methods and methods based on changes in measured T_2 relaxation are illustrated in Chap. 3. These may give more information on tissue perfusion but again are currently limited to measuring the flow in larger vessels. Spin enhancement agents also offer the possibility of measuring blood flow by methods similar to those used in CT and in nuclear medicine, and may allow the measurement of both perfusion and blood flow. The high contrast intrinsically present in blood in NMR images also allows cardiac function measurements with higher resolution than is possible in nuclear medicine (see Chap. 11).

Diffusion

It is known that the amplitude of the echoes following a 180° RF pulse is partially a function of the diffusion of the spins (Mansfield and Morris 1982). This self-diffusion coefficient is likely to vary between tissues and may therefore be of considerable interest. In attempting to measure the self-diffusion coefficient it is important to avoid also measuring the similar changes that occur as a result of perfusion in tissues and this presents a formidable problem for the measurement of self-diffusion. It has been possible to demonstrate perfusion in media which are not subject to blood flow, for instance the yolk of an egg (Taylor and Bushell 1985), but it is not clear whether it will also be possible to make these measurements in perfused tissue.

Paramagnetic Spin Enhancement Agents

These are molecules or atoms having unpaired electrons. They therefore have a large magnetic moment and produce a large local magnetic field. This field can produce enhanced T_1 and T_2 relaxation producing an increased z component and reduced x-y component of the net proton magnetic moment. The effect of this will depend on the spin sequence used but will produce increased signal when sampling z magnetisation (Runge et al. 1983; Brasch 1983; Crooks et al. 1982; Brown and Johnson 1984; Brasch et al. 1984). In spectroscopy these agents may cause chemical shifts and as a result of the increase in the T_2 relaxation they may also cause line broadening. They can be used as molecular probes in spectroscopy and in imaging can be used to obtain functional information similar to that obtained in nuclear medicine, but providing good localisation with high resolution and clear anatomical structure. Most of the agents currently under consideration, however, exhibit quite a high degree of toxicity to normal tissue and their value, therefore, depends upon stable binding to suitable compounds.

Other NMR Nuclei

After hydrogen, ^{31}P is currently the most important NMR nucleus for in vivo human measurements. [31]P spectroscopy provides information on the cellular energy cycle, giving information on conditions disturbing the phosphorus metabolism, both as a result of metabolic demands, and as a result of gross functional effects disturbing cell behaviour (Gadian 1982; Radda et al. 1983). A particular problem with spectroscopy is localising the origin of the spectroscopic signal to ensure that signal is only obtained from the region of interest. Aspects of this problem are discussed by Gordon (1985).

[13]C is used for spectroscopic studies of carbon compounds. However, due to the low natural abundance of the isotope ^{13}C it is usually necessary to use isotopically enriched compounds and whilst this is practicable, although relatively expensive, in animal studies, it will be very expensive in in vivo human studies (Algar and Shulman 1984). [19]F is of considerable interest in both spectroscopy and imaging. There are a number of blood-substitute compounds making use of fluorine and it can also be readily attached to a variety of compounds. It also gives a high signal compared with ^{31}P and [13]C (McFarland et al. 1985). [23]Na images are a valuable tool for investigating · the sodium balance in the brain, and also give information on the cellular sodium ion pump. Pathological changes resulting in an increase in the extracellular space often result in an increased sodium concentration in tissue (Maudsley and Hilal 1984; Hilal et al. 1985).

The use of an extracellular chemical shift agent such as dysprosium together with chemical shift imaging offers the possibility of separating the contributions of intra- and extracellular sodium.

Instrumentation

The Magnet

The magnet is central to NMR imaging and spectroscopy and must be designed to provide a magnetic field with a high homogeneity over the relevant volume, about

20 ppm for imaging and 0.1 ppm for spectroscopy. Low field systems are resistive and generally air, or more typically, water cooled, to dissipate the heat produced by the coils. At higher fields it becomes impractical to use resistive coils and superconducting coils are required. These need much more complex magnet design with heat shields and liquid helium and nitrogen jackets to maintain the coils in the superconducting state. However, because the current is locked into the coils they are, in principle, more stable and show only a very gradual decay in magnetic field due to a slight residual wire resistance. Permanent magnets, which are of great mass and have a limited maximum field strength, and electro-ferromagnetic magnets, which reduce the required current consumption for a given field strength, are also used in some NMR systems. Magnets for NMR application in medicine are discussed by Hanley (1984).

Shimming

Although magnets are carefully designed to have homogeneous magnetic fields over the imaging volume, fine adjustments to the field are required to obtain maximum homogeneity. These adjustments correct for slight departures in the final instrument from the design field and homogeneity, and also correct for the presence of environmental steel at the installation site. In spectroscopy systems the homogeneity required is such that the field must be readjusted for each sample, to remove the effects of the sample's magnetic susceptibility. These adjustments are normally made by sets of typically eight or thirteen shim coils which produce small compensating magnetic fields. For some systems much of the shimming may be carried out with sheets of steel dispersed about the magnet, resulting in reduced shim coil power consumption. Sometimes some of the shim coils are superconducting.

Gradient Coils

In order to carry out imaging it is necessary to provide magnetic field gradients and these are produced by independently driving three sets of coils to provide orthogonal gradients in the x, y and z directions. In order to localise in spectroscopy, gradient coils may also be used, although another approach is to use field profiling coils to provide an area of homogeneous field at only one point within the magnet.

RF Coils

The RF coils are used to transmit and receive RF signals which lie in the x-y plane with respect to the B_0 field. In their simplest form, they are therefore saddle shaped coils. In many machines, particularly at high field strength and frequencies, more complex coils based on cavity resonator principles are used. The machine normally has a body coil to transmit RF and to receive it from the large volume of the body. A head coil is used to provide a better signal for head images; as the signal-to-noise increases as the coil size decreases. Surface coils are used to provide high quality images of localised regions of the body, particularly the ear, eye, spine, neck and breast (Fisher et al. 1985). In these cases the body coil is normally used to transmit, which avoids high localised power deposition in the body and provides a more uniform image. However the sensitivity of the surface coil is not homogeneous and the non-uniformities will affect quantitative measurements. Spectroscopy also makes use of surface coils although here the coils are generally used for both transmission and reception. The coil serves to localise the region of interest and to provide optimum signal-to-noise.

RF Screening

In NMR measurements the signal being received is small and the frequency lies within radio broadcast wave bands. NMR is thus susceptible to interference from these sources and when transmitting, could itself be a source of interference. For this reason it is necessary to provide RF screening to the sensitive volume. This may be achieved by merely screening the bore of the magnet, although this can increase claustrophobia as it is necessary to ensure that the patient does not act as an aerial and introduce RF into the magnet. An alternative approach, which provides more flexibility in using the NMR system, and increased RF attenuation, is to screen the whole examination room.

Siting and Safety

Siting

The major consideration in siting an NMR machine is the effect of the magnet on equipment in its environment and the effect of the environment on the magnetic field homogeneity. The types of equipment affected by the magnet are primarily those in which electrons are accelerated over extended paths, where the electron's path will be distorted by the presence of the magnetic field, in photo-multiplier tubes on gamma cameras and other radiation counting equipment, in image intensifiers, linear accelerators and electron microscopes. The problems can be minimised by local magnetic shielding at the magnet or by local shielding of equipment or by producing an equal and opposite magnetic field by use of a coil placed around the equipment concerned. The degree to which a magnet will be affected by local constructional steel depends on the field strength and homogeneity required.

The most demanding situation is where a whole body magnet is to be used for spectroscopy, where the B_0 field must be some 200 times more homogeneous than is required for imaging. The effect of fixed steel work can often be compensated for by shimming or by dispersing suitable pieces of steel in mirror positions to counteract the effect of structural steel work. Mobile steel in doors, lifts or vehicles poses a much more serious problem as it cannot be easily compensated for. These causes of inhomogeneities must be avoided in siting and in designing facilities and many of these problems may be reduced by using a magnetically shielded magnet. Particular care must also be taken if the field strength is to be varied, as hysteresis in materials close to the magnet may cause the new induced field to be distributed differently to that occurring for the previous field setting.

Safety

Cardiac Pacemakers

Cardiac pacemakers are adversely affected by strong magnetic fields and also by alternating magnetic fields. The normal reaction is for the pacemaker to switch to automatic mode, producing a higher than normal pulse rate. However some pacemakers can be reprogrammed or stopped when exposed to high or changing magnetic fields. It is therefore imperative that both inside and outside the NMR building, people at risk are prevented from approaching too close to the field. 0.5 mT is recommended as being the maximum safe field. Outside a building it is necessary to erect fences with warning signs and before admission staff, patients and visitors should be informed of the hazards and asked whether they have a pacemaker.

Ferromagnetic Clips, Plates, etc.

Patients and staff working close to the magnet should be screened for the presence of ferromagnetic plates in the skull, and for any other post-operative clips and such persons should not be allowed to approach the magnet. This is a particular hazard with high field magnets.

Ferromagnetic Objects

Any unsecured ferromagnetic object will be accelerated towards the magnet, reaching a high speed in the magnet bore. Small sharp objects such as scissors could do great damage to patients or to staff: larger ferromagnetic objects can be attracted to the magnet with sufficient force to partially crush anyone between them and the magnet. Thus particular care has to be taken to exclude all ferromagnetic objects from the vicinity of the magnet and to devise methods of ensuring that such objects cannot be introduced, particularly when staff become familiar with the system and are therefore less alert to the hazards. The hazards are proportional to the magnet field strength. It should be noted that it is a major operation to turn off the field of a superconducting magnet and that an emergency shut-down requires quenching the system, with the released energy of the magnetic field being dissipated by boiling off much of the liquid helium. This can involve considerable expense and may involve some risk of damaging the superconducting coils of the magnet. During an emergency quench, the field is typically reduced by 50% in 10 s and 99% in 30 s. A permanent magnet, of course, cannot be turned off and it would therefore be far more difficult to deal with major incidents with such a device. The stray field, however, is much smaller, reducing the risk of such incidents.

Cryogens

With a cryogenic system, attention must be paid to the hazards of the extremely cold liquid helium and nitrogen. These liquids can cause severe burning, and soft tissues are rendered brittle if they approach these low temperatures. Gaseous helium and nitrogen, given off when filling the magnet, as the pipes and the magnet are cooled, and during spillages, can displace oxygen in the atmosphere, leading to the possibility of unconsciousness occurring due to insufficient inhaled oxygen. It is therefore necessary to allow provision for monitoring atmospheric oxygen levels and for flushing the air in the examination room in the event of a large spillage or leak of cryogen. Gas boiled off from the magnet should usually be vented to the atmosphere via a suitable pipe.

Superconducting Magnet Quench

A superconducting magnet can be subject to a spontaneous quench, resulting in the release of considerable quantities of helium gas. As noted above, if vented to the examination room, this can result in both potential overpressure and asphyxiation. Provision is usually made to vent any released gases to the atmosphere via a large diameter vent pipe. An alternative approach (Bore and Timms 1984) is to design the magnet room so that the low density helium released during a quench will not occupy the lower 2 m of the room, and also to provide the room with windows or panels designed to open on overpressure. This latter design criterion could be difficult to achieve in an RF screened room.

Biological Effect of NMR Scanning

Patients undergoing NMR scan, and to some extent staff conducting NMR examinations, are exposed to static magnetic fields, time varying magnetic fields, and RF electromagnetic fields. Several authors have considered the possible physiological effects or long term hazards that might arise from NMR exposure (Saunders and Smith 1984; Budinger 1979; NRPB 1983) with the NRPB Advisory Group producing guidelines for NMR clinical measurements. In the USA, the FDA has also produced safety limits (FDA 1982).

Experimental evidence of physiological or mutagenic effects arising from static magnetic field exposure is contradictory and there is an absence of clear, corroborated findings. It is known that moving electrolytes in the presence of a magnetic field will result in an induced current and potential difference being generated. At magnetic fields in excess of 2.5 T, the potential is approaching that required to directly stimulate normal cardiac cells. Thus the NRPB have limited static magnetic field exposure to 2.5 T. Differences in electrocardiograms have been reported, but this is attributed to interference in signal recording due to potential differences induced by flow effects, rather than to changes in cardiac behaviour.

Some effects due to alternating fields have been reported, again generally due to induced current or potential, and limits have been suggested on the basis of the induced current density required to produce ventricular fibrillation in normal tissue. The NRPB have recommended that the z gradient magnetic flux density should not exceed an rms rate of change of 20 Ts^{-1} for pulses of length 10 ms.

Radiofrequency fields are known to produce local heating if sufficient RF power is absorbed. At low power deposition, resulting in temperature increases of no more than 1°C, there is good evidence of adverse effects, and the NRPB guideline is that the whole body should not absorb RF power in excess of 0.4 wkg^{-1}.

References

Algar JR, Shulman RG (1984) Metabolic applications of high-resolution ^{13}C nuclear magnetic resonance spectroscopy. Br Med Bull 40: 160–164

Andrew ER, Bottomley PA, Hinshaw WS et al. (1977) NMR images by the multiple sensitive point method: Application to larger biological systems. Phys Med Biol 22: 971–974

Bendall MR, Gordon RE (1983) Depth and refocussing pulses designed for multipulse NMR with surface coils. J Magn Reson 53: 365

Bore PJ, Timms WE (1984) The installation of high-field NMR equipment in a hospital environment. Magn Reson Med 1: 387–395

Bottomley PA, Foster TB, Darrow RT (1984) Depth-resolved surface-coil spectroscopy (DRESS) for in vivo ^{1}H, ^{31}P and ^{13}C NMR. J Magn Reson 59: 338–342

Bottomley PA, Smith LS, Leue WM, Charles C (1985) Slice interleaved depth resolved surface-coil spectroscopy (SLIT DRESS) for rapid ^{31}P NMR in vivo. Proceedings of the 4th Meeting of the Society of Magnetic Resonance in Medicine. SMRM, Berkeley, CA, pp 946–947

Bradley WG, Waluch V (1985) Blood flow: Magnetic resonance imaging. Radiology 154: 443–450

Brasch RC (1983) Work in progress: methods of contrast enhancement for NMR imaging and potential applications. Radiology 147: 781–788

Brasch RC, Weinman HJ, Websey GE (1984) Contrast enhanced NMR imaging: animal studies using gadolinium-DPTA complex. AJR 142: 625–630

Brown MA, Johnson GA (1984) Transition metal-chelate complexes as relaxation modifiers in nuclear magnetic resonance. Med Phys 11: 67–72

Brunner P, Ernst RR (1979) Sensitivity and performance time in NMR imaging. J Magn Reson 33: 83–106

Budinger TF (1979) Thresholds for physiological effects due to RF and magnetic fields used in NMR imaging. IEEE. Trans Nucl Sci N526: 2821–2825

Carr HY, Purcell EM (1954) Effects of diffusion in free precession in nuclear magnetic resonance expts. Phys Rev 94: 630

Cox SJ, Styles P (1980) Towards biochemical imaging. J Magn Reson 40: 209–212

Crooks LE, Kaufman L (1984) NMR imaging of blood flow. Br Med Bull 40: 167–169

Crooks LE, Mills CM, Davis PL et al. (1982) Visualisation of cerebral and vascular abnormalities by NMR imaging. The effect of imaging on contrast. Radiology 144: 843–852

Damadian R, Minkoff L, Goldsmith M et al. (1976) Field focussing nuclear magnetic resonance (FONAR): Visualisation of a tumour in a live animal. Science 194: 1430–1432

Dixon WT (1983) Chemical shift imaging with proton NMR. Radiology 149(P): 238

Edelstein WA, Hutchison JMS, Johnson G, Redpath T (1980) Spin warp. NMR imaging and applications to human whole-body imaging. Phys Med Biol 25: 751–756

FDA (1982) United States Bureau of Radiological Health of the Food and Drug Administration (BRM (HFX-460), FDA, 12 February 1982)

Fisher MR, Barker B, Amparo EG et al. (1985) MR Imaging using specialized coils. Radiology 157: 443–447

Gadian DG (1982) Nuclear magnetic resonance and its applications to living systems. Oxford University Press, Oxford

George CR, Jacobs G, MacIntyre WJ et al. (1984) Magnetic resonance signal intensity patterns obtained from continuous and pulsatile flow models. Radiology 151: 421–428

Gordon RE (1985) Magnets, molecules and medicine. Phys Med Biol 30: 741–770

Gordon RE, Hanley PE, Shaw D et al. (1980) Localisation of metabolites in animals using ^{31}P topical magnetic resonance. Nature 287: 736–738

Haase A, Frahn J, Matthaei D (1985) Multi-line chemical shift selective (MULTI-CHESS) imaging using stimulated echoes. Proceedings of the 4th meeting of the Society of Magnetic Resonance in Medicine. SMRM, Berkeley, CA, pp 155–156

Hanley P (1984) Magnets for medical applications of NMR. Br Med Bull 40: 125–131

Hilal SK, Maudsley AA, Ra JB et al. (1985). In vivo NMR imaging of sodium-23 in the human head. J Comput Assist Tomogr 9: 1–7

Hinshaw WS (1976) Image formation by NMR: The sensitive point method. J Appl Phys 47: 3709–3721

Hoult DI (1979) Rotating frame zeumatography. J Magn Reson 33: 183

Hounsfield GN (1973) Computerised transverse axial scanning (tomography). Part 1. Description of system. Br J Radiol 46: 1016–1022

King KF, Moran PR (1984) A unified description of NMR imaging, data collection strategies, and reconstruction. Med Phys 11: 1–14

Kumar A, Welti D, Ernst RR (1975a) Imaging of microscopic objects by NMR Fourier zeumatography, Naturwissenschaften 62: 34

Kumar A, Welti D, Ernst RR (1975b) NMR Fourier zeumatography, J Magn Reson 18: 69

Lai C-M (1981) True three-dimensional nuclear magnetic resonance imaging by Fourier reconstruction zoomatography. J Appl Phys 52: 1141–1145

Lauterbur PC (1973) Image formation by induced local interactions: Examples employing nuclear magnetic resonance. Nature 242: 190

Leach MO, Hind AJ, Sauter R, Requardt H, Weber H (1985) Design and use of a dual-frequency surface coil providing proton images for improved localistion in ^{31}P spectroscopy of small lesions. Radiology 157(P): 319

McFarland E, Koutcher JA, Rosen BR, Telcher B, Brady TJ (1985) In vivo ^{19}F NMR imaging. J Compt Assist Tomogr 9: 8–15

Mansfield P (1977) Multi-planar image formation using NMR spin echoes. J Phys C 10: 225–258

Mansfield P (1981) Critical evaluation of NMR imaging techniques. Proc Int Symposium on Nuclear Magnetic Resonance Imaging. Bowman Gray School of Medicine, Winston-Salem

Mansfield P, Morris PG (1982) NMR imaging in bio-medicine. Academic Press, London

Mansfield P, Pykett IL (1978) Biological and medical imaging by NMR. J Magn Res 29: 355–373

Mansfield P, Pykett IL, Morris PG, Coupland RE (1978) Human whole-body line-scan imaging by NMR. Br J Radiol 51: 921–922

Maudsley AA, Hilal SK (1984) Biological aspects of sodium-23 imaging. Br Med Bull 40: 165–166

Maudsley AA, Hilal SK, Perman WH, Simon HE (1983) Spatially resolved high resolution spectroscopy by "four-dimensional" NMR. J Magn Reson 51: 147–152

Meiboom S, Gill D (1958) Measuring nuclear relaxation times. Rev Sci Instr 29: 688

Mills CM, Brant-Zawadzhi M, Crooks LE et al. (1983) Nuclear magnetic resonance: Principles of blood flow imaging. AJNR 4: 1161–1166

Moran PR, Moran RA, Karstaedt N (1985) Verification and evaluation of internal flow and motion. Radiology 154: 433–441

NRPB (1983) Revised guidance on acceptable limits of exposure during nuclear magnetic resonance clinical imaging. NRPB ad hoc advisory group on NMR clinical imaging. Br J Radiol 56: 974–977

O'Donnell M (1985) NMR blood flow imaging using multiecho, phase contrast sequences. Med Phys 12: 59–64

Ordridge RJ (1985) Localised chemical shift measurements in phosphorus and protons. Proceedings of the 4th Meeting of the Society of Magnetic Resonance in Medicine. SMRM, Berkeley, CA, pp 131–132

Radda GK, Bore PT, Rajagopalan B (1983) Clinical aspects of ^{31}P NMR spectroscopy. Br Med Bull 40: 155–159

Runge VM, Clanton JA, Lukehart CM, Partan CM, James AE (1983) Paramagnetic agents for contrast-enhanced NMR imaging: a review. AJR 141: 1209–1215

Saunders RD, Smith H (1984) Safety aspects of NMR clinical imaging. Br Med Bull 40: 148–154

Singer JR (1978) NMR diffusion and flow measurements and an introduction to spin phase graphing. J Phys E Sci Instrum 11: 281–291

Taylor DG, Bushell MC (1985) The spatial mapping of translational diffusion coefficients by the NMR imaging technique. Phys Med Biol 30: 345–349

Bibliography

Books and Volumes

Physics

Abragam A (1983) Principles of nuclear magnetism. Oxford University Press, Oxford

Andrew ER (1958) Nuclear magnetic resonance. Cambridge University Press, Cambridge

Mansfield P, Morris PG (1982) NMR imaging in bio-medicine. Academic Press, London

General

Beall PT, Amtey SR, Kasturi SR (1984) NMR data handbook for biomedical application. Pergamon Press, Oxford

British Medical Bulletin (1984) Volume 40 No. 2

Esser PD, Johnston RE (1984) Technology of nuclear magnetic resonance. Society of Nuclear Medicine, New York

Farrer TC, Becker ED (1971) Pulse and Fourier transform NMR: introduction to theory and methods. Academic Press, New York.

Foster MA (1984) Magnetic resonance in medicine and biology. Pergamon Press. Oxford

Gadian DG (1982) Nuclear magnetic resonance and its applications to living systems. Oxford University Press, Oxford

Hopf MA, Smith FW (1984) Magnetic resonance in medicine and biology. Karger, Basel

Kaufman L, Crooks LE, Margulis AR (1981) Nuclear magnetic resonance (NMR) in medicine. Igaku-Shoin, New York

Lerski RA (1985) Physical principles and clinical applications of nuclear magnetic resonance. Institute of Physical Sciences in Medicine, London

Partain CL, James AE, Rollo FD, Price RR (1983) Nuclear magnetic resonance (NMR) imaging. W. B. Saunders, Philadelphia

Philosophical Transactions of the Royal Society of London (1980) 289 (1037) The Royal Society, London

Podo F, Orr JS (1983) In: Annali dell'Istituto Superiore di Sanita. EEC workshop on identification and characterisation of biological tumours by NMR. Vol. 19 No. 1

Proceedings of the Society of Magnetic Resonance in Medicine Annual Meetings (New York 1984, London 1985, Montreal 1986) SMRM, Berkeley, CA.

Shaw D (1976) Fourier transform NMR spectroscopy. Elsevier, Amsterdam

Slichter CP (1978) Principles of magnetic resonance. Springer, Berlin, Heidelberg, New York

Witcofski RL, Karstaedt N, Partain CL (1982) NMR imaging. Proceedings of an International Symposium in Nuclear Magnetic Resonance Imaging. Bowman Gray School of Medicine, Winston-Salem, NC

Other Introductory and Review Articles

Bradley WG (1982) NMR tomography. Diasonics, Milpitas, California

Budinger TF (1981) Nuclear magnetic resonance (NMR) in vivo studies: known threshold for health effects. J Comput Assist Tomogr 5: 800–811

Budinger TF, Lauterbur PC (1984) Nuclear magnetic resonance technology for medical studies. Science 226: 228–298

Dixon RL, Elstrand KE (1982) The physics of proton NMR. Med Phys 9: 807–818

FDA (1982) Guidelines for evaluating electromagnetic risks for trials of clinical NMR systems. BRM (HFX-460) (Bureau of Radiological Health of the USA Food and Drug Administration) February 12 1982

Goni FM (1984) NMR: Principles and biological applications. In: Chayen T, Bitensky L, (eds) Investigative techniques in medicine and biology, vol 1. Mercel Dekker, New York, pp 1–39

Gordon RE (1985) Magnets, molecules and medicine. Phys Med Biol 30: 741–770

Gore JC, Emery EW, Orr JS, Doyle FH (1981) Medical nuclear magnetic resonance imaging: 1. Physical principles. Invest Radiol 16: 269–274

Gore JC (1983) Physical principles of nuclear magnetic resonance imaging. In: Steiner RE (ed) Recent advances in radiology and medical imaging. Churchill Livingstone, Edinburgh, pp 1–13

Hinshaw WS, Lent AM (1983) An introduction to NMR imaging: from the block equation to the imaging equations. Proceedings of the IEEE 71 (3): 338–350

Hoult DI (1980) Medical imaging by NMR. In: Cohen JS (ed) Magnetic resonance in biology. John Wiley, New York, pp 70–109

King KF, Moran PR (1984) A unified description of NMR imaging, data collection strategies and reconstruction. Med Phys 11 (1): 1–14

NRPB Ad Hoc Advisory Group on NMR Clinical Imaging (1983) Revised guidance on acceptable limits of exposure during nuclear magnetic resonance clinical imaging. Br J Radiol 56: 974–977

Pykett IL (1982) NMR imaging in medicine. Sci Am 246 (5): 78–88

Pykett IL, Newhouse JH, Buonanno FS et al. (1982) Principles of nuclear magnetic resonance imaging. Radiology 143: 157–168

Pykett IL, Buonanno FS, Brady TJ, Kistler JP (1983) Techniques and approaches to proton NMR imaging of the head. Comput Radiol 7 (1): 1–17

3. Functional Imaging Using NMR

R. Bachus, E. Mueller, H. Koenig, G. Braeckle, H. Weber
and E.R. Reinhardt

NMR Image Parameters

The signal intensity and contrast of NMR images depend both on measurement parameters and on tissue parameters. The most essential measurement parameters are the repetition time TR and the spin echo time TE. The effect of the echo time on the image contrast is demonstrated by the multi echo images shown in Fig. 3.1. The significant change in image contrast with increasing echo time is clearly demonstrated. The tissue specific parameters are the relaxation times T_1 and T_2, the proton density P and the diffusion coefficient D. Furthermore, the signal intensity in vessels is also fundamentally influenced by the flow velocity of the blood (see below). By applying chemical shift imaging methods, it is also possible to generate separate images of fat, water and magnetic susceptibility.

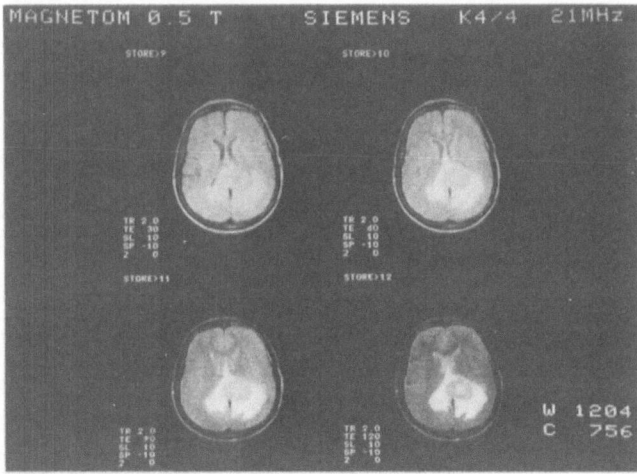

Fig. 3.1. The first four spin echoes of a hypernephroma metastasis obtained by an interleaved pulse sequence.

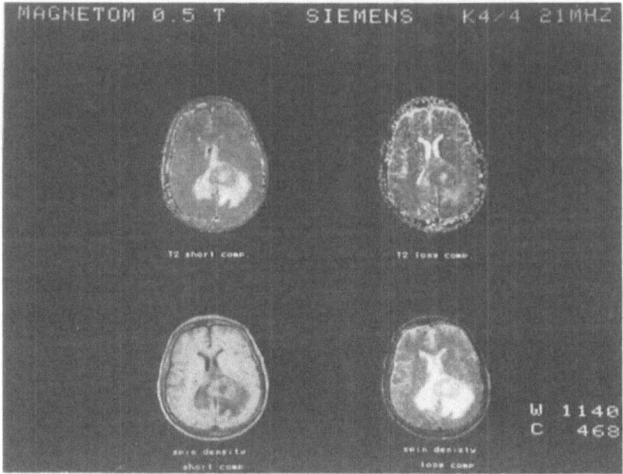

Fig. 3.2.   Short and long component T_2 and corresponding spin density.

T_1, T_2, P and Multi-exponential Processes

The image parameters T_1, T_2 and P can be computed pixelwise from a suitable set of images. A pixel in the NMR image corresponds to a volume element in the body. Within any such volume element the protons are bound under a range of different chemical conditions and are therefore likely to be characterised by more than one T_2 relaxation time per pixel. Assuming two different exponential processes, we computed for each pixel in the set of images in Fig. 3.1 two T_2 values together with corresponding T_1 weighted spin density values. Figure 3.2 shows the four images resulting from this analysis. The short T_2 component emphasises the oedema and the long T_2 component the CSF. Differences in these two relaxation times reflect different biochemical conditions. Each T_2 image and its corresponding "spin density" image provide different views of normal tissue and tumour tissue. A T_1 image can be calculated if the same slice is imaged with a different repetition time, as shown in Fig. 3.3 for the same data.

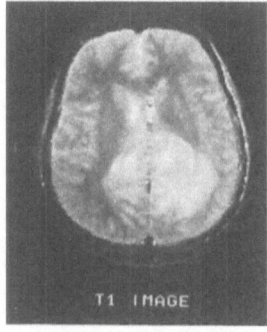

Fig. 3.3. T_1 image.

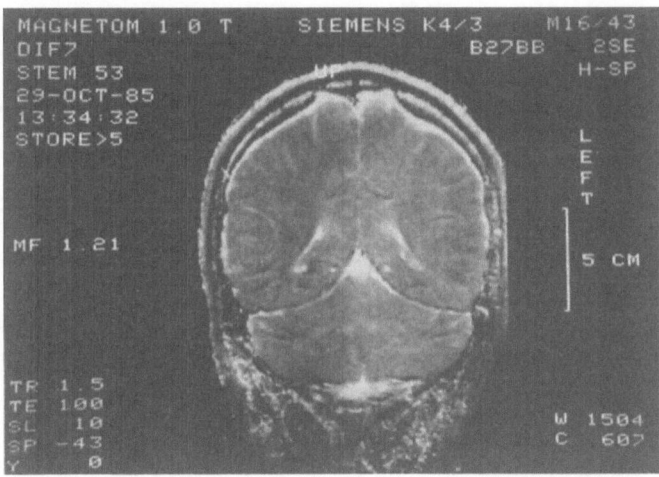

Fig. 3.4. Diffusion image

Diffusion Coefficient Images

Another parameter characterising tissue properties is the diffusion coefficient D. D can be measured by using a spin echo pulse sequence with additional gradient pulses (Stejskal and Tanner 1965). The dependence of the spin echo amplitude E on the gradient pulses is given by the relationship:

$$E = E_o . \exp\left(-\gamma^2 g^2 \delta^2 (\Delta - \delta/3) . D\right) \quad (1)$$

where γ denotes the gyromagnetic ratio for protons, g and δ denote the amplitude and width of the gradient pulses respectively, and Δ is the delay between gradient pulses. E_o denotes the echo amplitude without gradient pulses.

Therefore if two spin echo images are measured using different gradient amplitudes the diffusion image can be calculated, as shown in Fig. 3.4 for a coronal section through the head. In living tissue, the diffusion image will also include a contribution from tissue perfusion.

Chemical Shift and Susceptibility Imaging

The chemical shift between CH_2 and water protons can be used to generate separate images of fat and water. It is assumed that the increased information available will lead to improved diagnosis in conditions with abnormal fat water ratios. It is also possible to calculate the NMR relaxation times T_1 and T_2 for both components. The image quality is also improved due to the suppression of the chemical shift artefact caused by the spatial displacement between fat and water in non-corrected images. Several methods have been developed for fat/water separation. The technique proposed by Dixon (1984) requires two measurements with different phases of the 180° RF pulse with respect to the gradients in the spin echo sequence. An extension of this

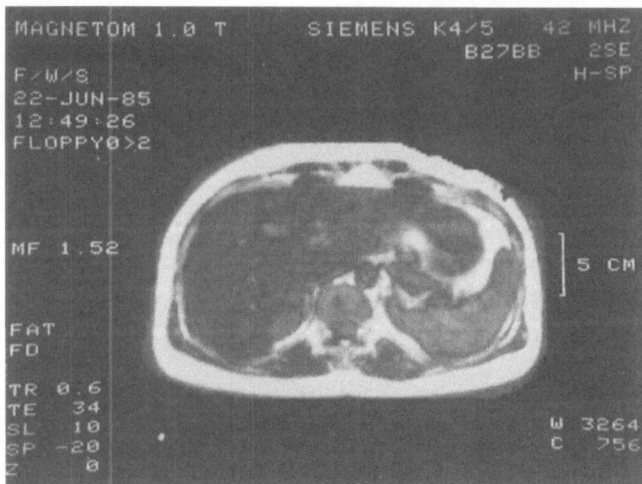

Fig. 3.5. Fat image, obtained with the Dixon method.

method allows the calculation of the magnetic susceptibility (Margosian 1985). Examples are given in Figs. 3.5, 3.6 and 3.7 respectively for a transverse section through the thorax. An alternative approach, the CHESS method (Rosen et al. 1984; Haase et al. 1985) eliminates one component by the application of a frequency selective 90° RF pulse and a subsequent de-phasing gradient. After this preparative procedure the remaining component can be measured by using standard imaging sequences. Figures 3.8 and 3.9 show fat and water images of a hip joint. The cartilage is clearly visualised in the water image.

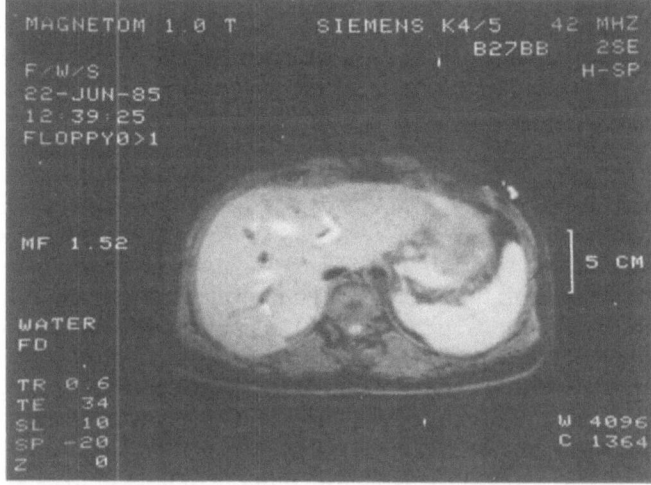

Fig. 3.6. Water image, obtained with the Dixon method.

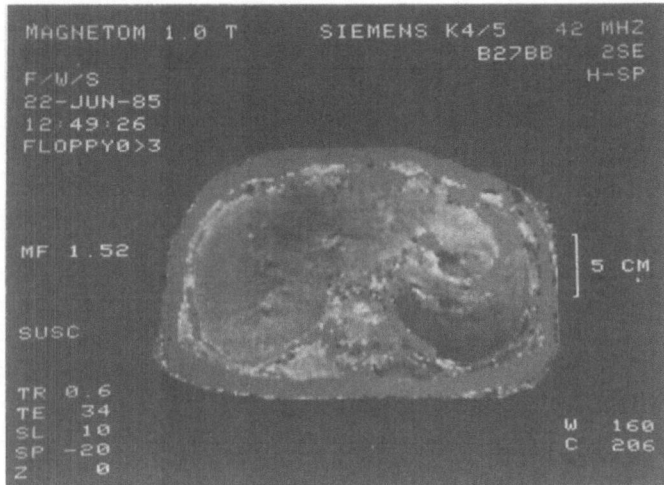

Fig. 3.7. Susceptibility image.

Contrast Optimised Images

The large number of potential NMR images that can be obtained at a given site, resulting from the range of possible imaging parameters, leads to the following questions: Which image parameters contain the most useful diagnostic information? Is it possible to compress all of the available information into a few contrast optimised images to provide images of improved diagnostic content? Is it possible to improve the specificity of NMR images for clinical diagnosis?

Development of a "Training" Database

These questions can only be examined systematically if a database of representative labelled images is available. This means that regions in the NMR image containing different tissue classes have to be identified by a radiologist who knows the true diagnosis. Some appropriate labels are: normal tissue, tumour, oedema and cerebro-spinal fluid. This information needs to be derived by histology or by the gold standard procedure. The labelling procedure has to be applied to a number of representative cases in order to create the database.

Experimental Results Obtained with a Trained Polynomial Classifier

A series of cases in a brain study have been measured using an interleaved pulse sequence which combined two different repetition times, a long TR with 16 echoes

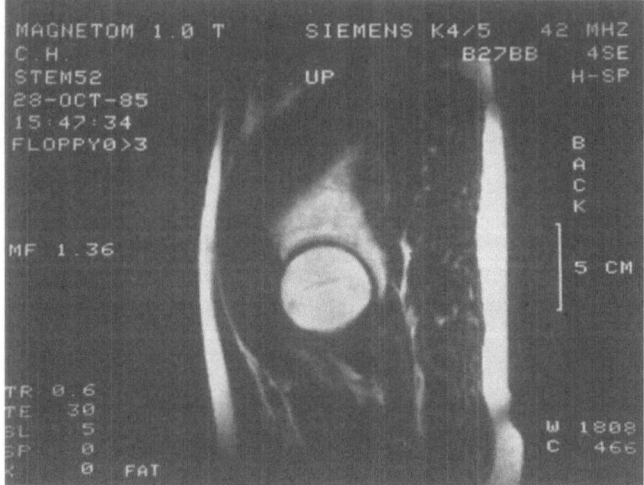

Fig. 3.8. Fat image of a hip joint, using the CHESS technique.

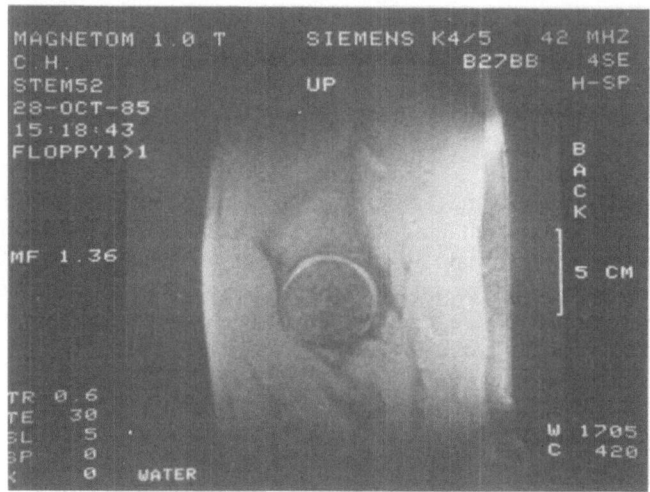

Fig. 3.9. Water image of the hip joint shown in Fig. 3.8 using the CHESS technique.

and a short TR with one echo. The following features have been derived from the NMR images:

Image intensity values
NMR parameters
Morphological features

The evaluation process is divided into training and working phases: during the training phase, feature vectors of randomly selected pixels are calculated from the labelled training set. These feature vectors, which contain information about the texture of the labelled regions, are used to train a linear polynomial classifier. Consequently this classifier is optimised to differentiate between the tissue classes of interest. In the working phase the classifier is applied to unknown tissues, leading to an estimate of the probability of class membership of a specific tissue type for each pixel. These probabilities are then displayed in new synthesised images, providing the optimum contrast for a given task. Figure 3.10 shows the classification result for the hypernephroma metastasis displayed in Fig. 3.1. As is evident, it is possible to differentiate between normal tissue, CSF, tumour tissue and oedema. Each recognised tissue class is represented by a different grey scale value in the image.

Another interesting and more detailed representation of these results are images in which the estimated probabilities for the different tissue classes are represented on a grey scale as shown in Fig. 3.11. High intensity corresponds to high probability for the particular tissue class. Pixels with very low probability are suppressed and represented with their grey scale value from the normal spin echo image.

Flow Imaging

The NMR signal depends not only on pulse sequence, spin density and relaxation times T_1 and T_2, but also on the velocity of the moving spins themselves. This gives moving blood an intrinsically high contrast and permits blood flow to be visualised non-invasively without contrast agents. Even small vessels in the brain can be discerned using an appropriate choice of measurement parameters (Fig. 3.12). The signal intensities of vessels depends very strongly on the delay time to the R wave in ECG gated images. The subtraction image of two images with different delays looks similar to a DSA-image (Fig. 3.13).

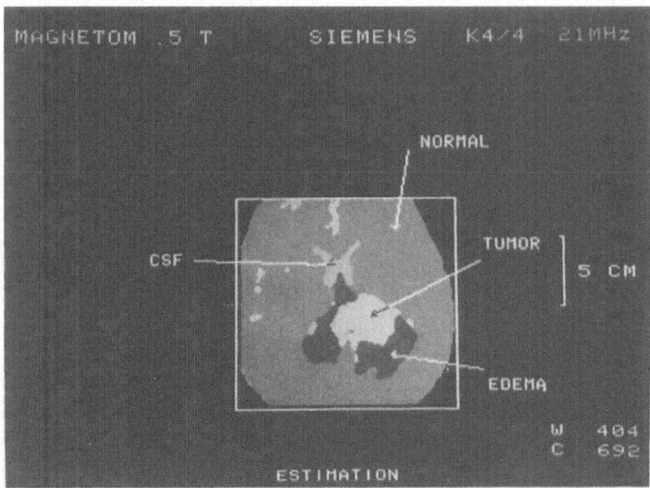

Fig. 3.10. Result of the classification process for the pathology shown in Fig. 3.1. The estimated class regions are sorted by intensity: oedema, normal tissue, CSF, tumour.

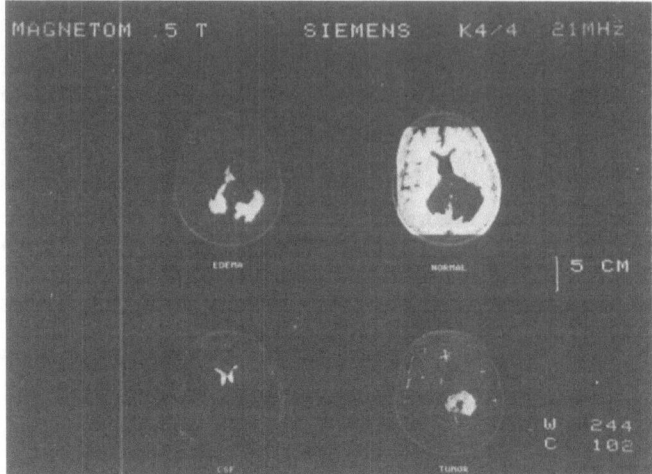

Fig. 3.11. Spin echo images with the probabilities represented on a grey scale for the four tissue types. High intensity corresponds to high probability.

Another possible method of enhancing vascular structures is to calculate phase images which are highly sensitive to motion. The NMR image is obtained by a two-dimensional Fourier transform of the complex time domain. In the case of stationary protons the image can be described completely by the real part whereas for moving protons both real and imaginary parts have to be taken into account. By using Eq. (2) a phase value (ϕ) that is essentially proportional to the flow velocity v can be computed pixel by pixel.

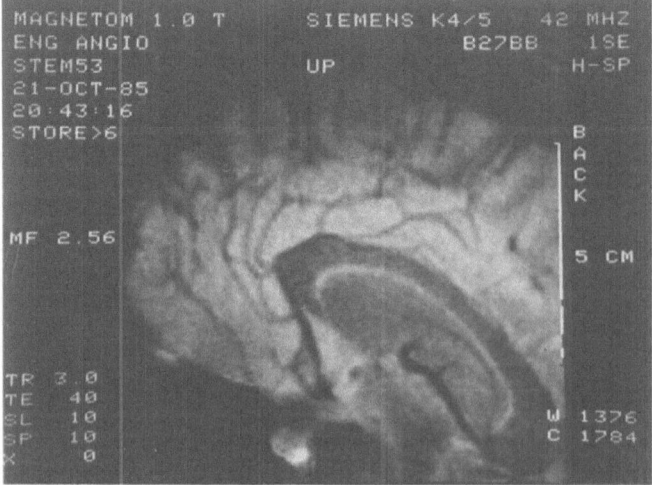

Fig. 3.12. Head image with good contrast between brain tissue and vessels.

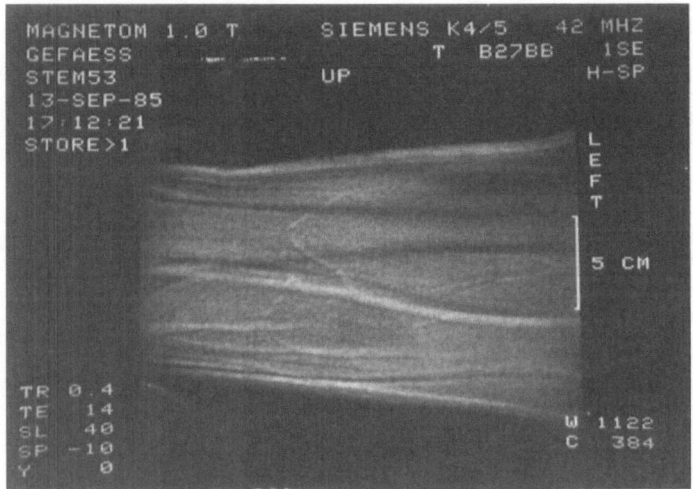

Fig. 3.13. Subtraction image of two images with different delays relative to the R wave. Even small arteries can be displayed by this technique.

$$v \sim \emptyset = \arctan \left\{ \frac{\text{imaginary part}}{\text{real part}} \right\} \tag{2}$$

The diagnostic value of this technique is demonstrated by Figs. 3.14 and 3.15. Figure 3.14 shows a patient with a large aortic aneurysm. On the basis of this spin echo image it is impossible to decide whether moving blood is present or whether there is already thrombotic tissue. The corresponding phase image provides additional information (Fig. 3.15). Moving blood is clearly indicated by low grey values in the region of the aneurysm.

Flow velocities can also be measured accurately by measuring the apparent relaxation times resulting from moving spins. For stationary fluid the NMR signal decays with a time constant T_2. When flowing spins pass through the selected slice the inflow and outflow of magnetisation also have to be considered. In simple form, the transverse relaxation rate $1/T_2$ can be modified according to Eq. (3) (Zhernovol 1965):

$$\frac{1}{T_{2f}} = \frac{1}{T_2} + \frac{v}{d} \tag{3}$$

where T_{2f} = measured relaxation time (apparent relaxation time)
T_2 = relaxation time of stationary fluid
v = velocity of spins
d = slice thickness

The validity of this equation has been verified by performing multi-echo experiments using a flow phantom and varying the ratio of v/d. The dependence of the transverse relaxation time on flow velocity, and on slice thickness in the case of stationary flow, is shown in Figs. 3.16 and 3.17 respectively.

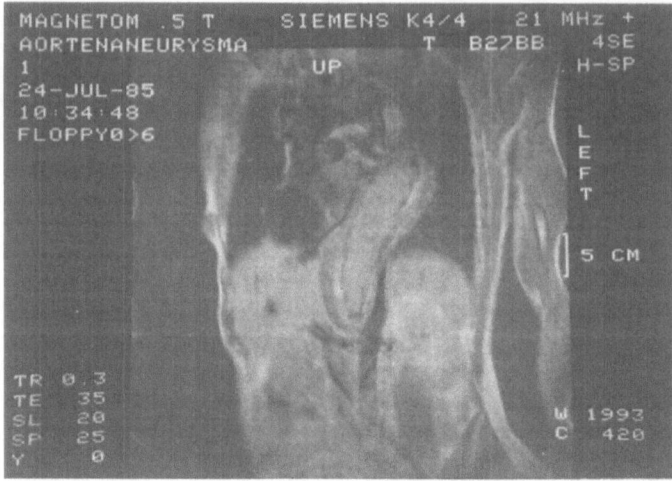

Fig. 3.14. Patient with a large aortic aneurysm.

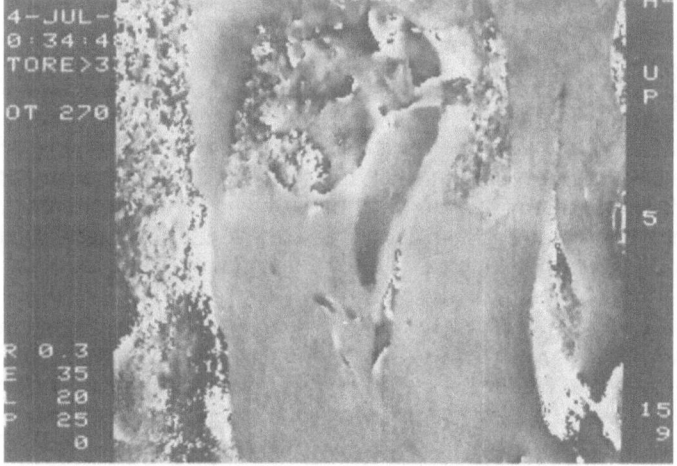

Fig. 3.15. Phase image for the pathology displayed in Fig. 3.14.

For pulsatile flow the signal intensity is a function of the instantaneous flow velocity in gated images. Figure 3.18 shows the integrated signal intensity in the aorta of a volunteer versus echo time obtained by two multi-echo experiments. The data points were measured with a trigger delay of 40 ms and 400 ms from the R wave, respectively. The modulation of the signal intensity with increasing time reflects directly the changing velocity of blood in this vessel. The actual flow velocity was calculated by Eq. (2) using four adjacent echoes (Mueller et al. 1986). Figure 3.19 shows the result of this analysis for a region of interest in the abdominal aorta. A significant reduction in the maximum velocity and a broadening of the systolic peak are

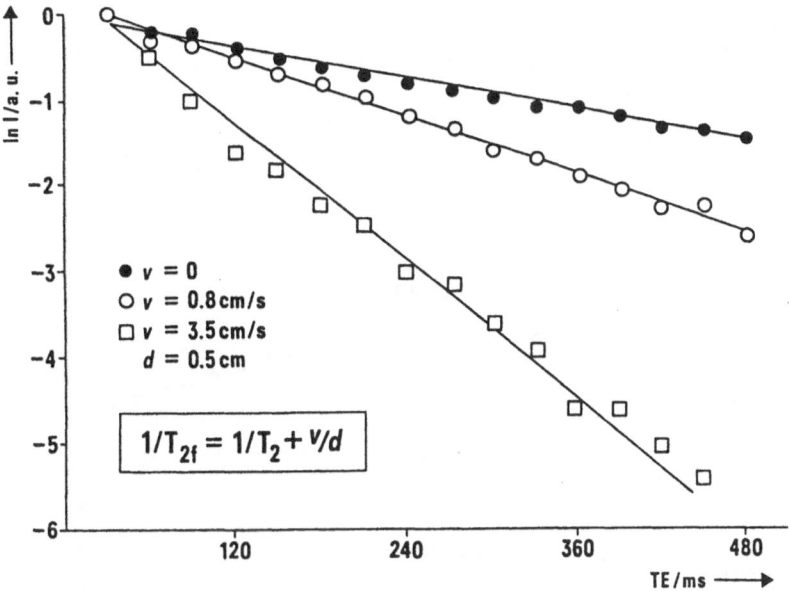

Fig. 3.16. Measured relaxation rates $1/T_{2f}$ of fluids flowing with different velocities v through a slice of thickness d.

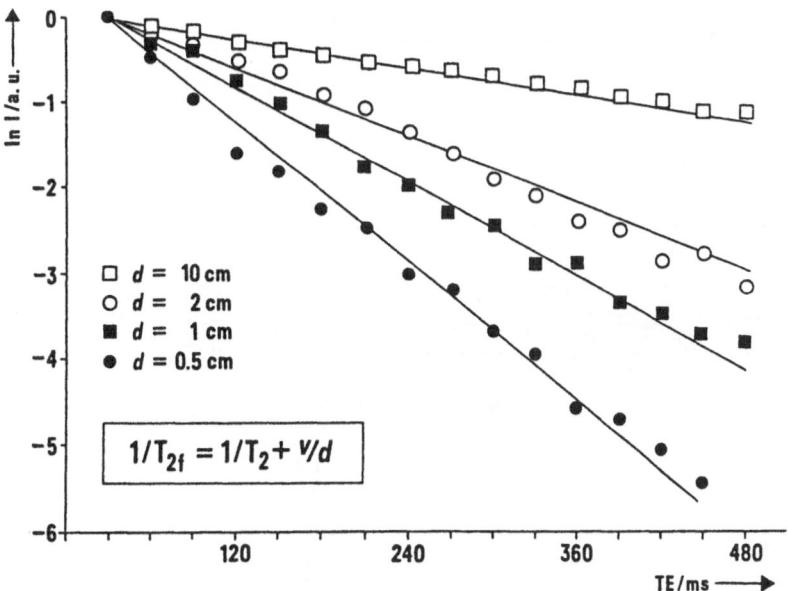

Fig. 3.17. Measured relaxation rates $1/T_{2f}$ of a flowing fluid ($v = 3.5$ cm/s) as a function of the slice thickness d.

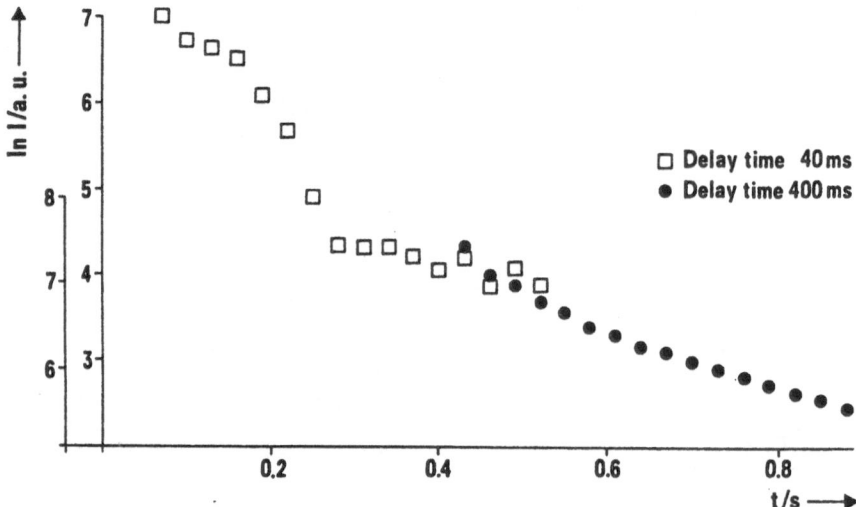

Fig. 3.18. ECG gated multiecho experiments at different delay times after the R wave in the abdominal aorta of a normal volunteer.

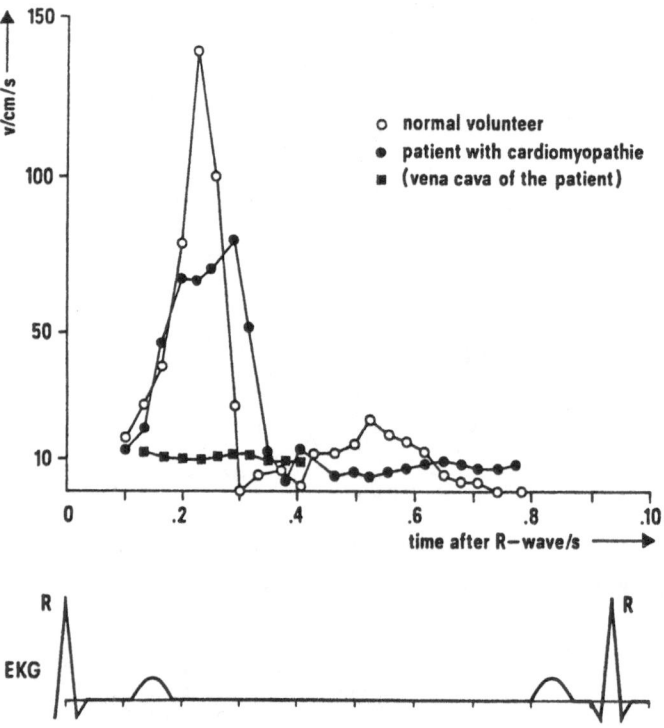

Fig. 3.19. Comparison of the pulsatile flow behaviour in the abdominal aorta between a normal volunteer and a patient with a cardiomyopathy.

observed for a patient with a cardiomyopathy in comparison to a normal volunteer. The results have been confirmed quantitatively by Doppler ultrasound measurements.

Cardiac Imaging

NMR techniques can be applied to cardiac investigations in order to improve the assessment of heart disease. ECG gating must be used to obtain images at selected phases of the cardiac cycle. The organ contours and the delineation of the myocardial tissue are of high quality compared with conventional imaging methods in cardiology (Fig. 3.20). The geometry and motion of the left ventricle can be evaluated by a

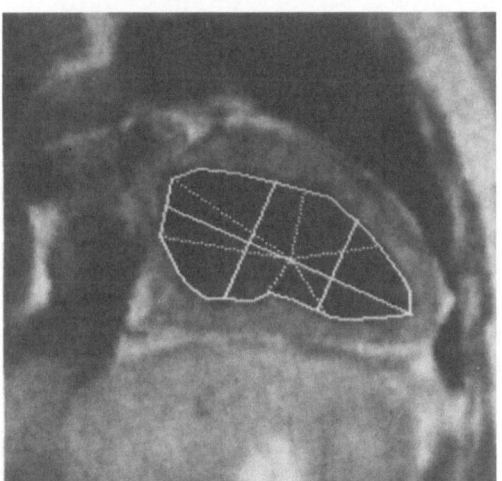

Fig. 3.20. Ventricular contour and regions at end diastole.

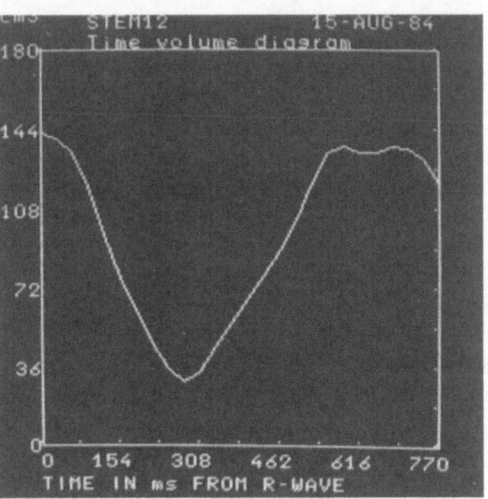

Fig. 3.21. Time volume diagram of the left ventricle.

sequence of images obtained at different phases of the heart cycle. The methods originally developed for use in nuclear medicine and in angiography have been transferred to NMR. Some of the parameters which can be derived include chamber volume as a function of the cardiac cycle (Fig. 3.21) and the ejection fraction.

As a result of being able to precisely define the inner and outer borders of the myocardium, the wall thickness and wall motion as a function of position or wall region can be directly displayed and measured. Figure 3.22 shows the variation in regional myocardial wall thickness for a volunteer at end systole (upper curve) and end diastole (lower curve). It is often difficult to accurately position the projection views in angiography. Parts of the outflow tract can be included and this sometimes

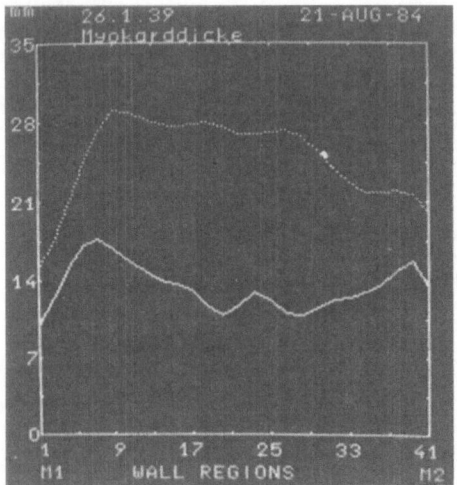

Fig. 3.22. Regional wall thickness along the circumference of the left ventricle at systole (*upper curve*) and diastole.

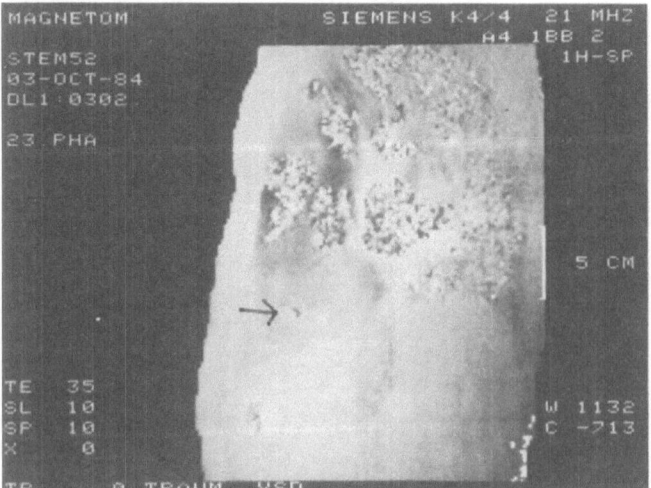

Fig. 3.23. Phase image of a traumatic ventricular septum defect. Flow in the shunt region.

leads to larger volumes than those obtained with NMR. The accuracy of NMR volume measurements can be further improved by measuring a second orthogonal slice, or by multislice imaging or direct 3D acquisition.

Phase images can be calculated in order to enhance visualisation of intracardiac flow phenomena. Figure 3.23 shows a traumatic injury of the septum which can be clearly identified in the phase image. Phase images can also be used to enable quantitative evaluation of dynamic processes. The phase value of a pixel is a function of the velocity component of that pixel in the readout direction. The velocity component perpendicular to the first velocity component can be determined from a second measurement with interchanged readout and phase-encoding gradients. Both results can then be combined to provide motion vectors indicating the direction and magnitude of the motion of heart segments at a particular time in the cardiac cycle (Fig. 3.24). The multislice capability of NMR can be used to generate three-dimensional displays. The acquisition of n slices each at n different times in the

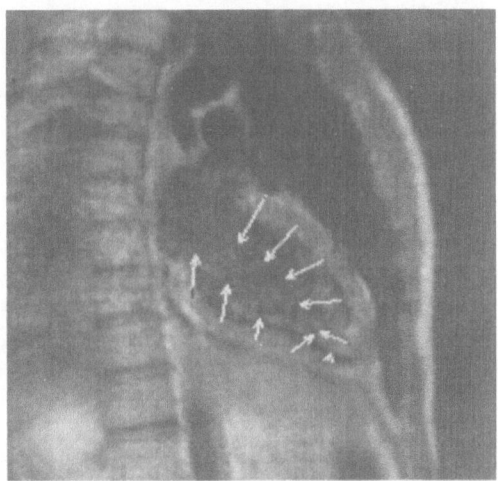

Fig. 3.24. Motion vectors from a reconstructed phase image. End systolic velocity vectors.

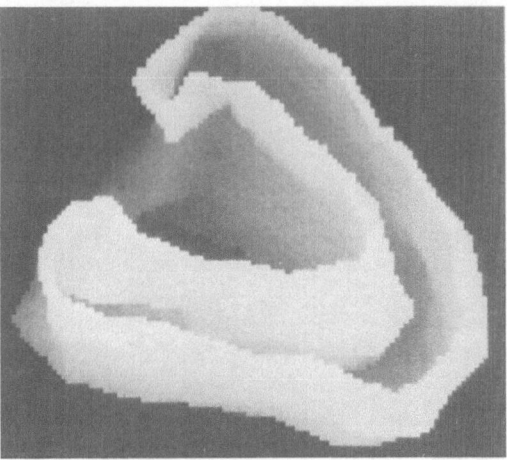

Fig. 3.25. 3D-reconstruction of the left ventricle from several slices.

cardiac cycle is required to produce a complete three-dimensional reconstruction of the left ventricle. As an example, Fig. 3.25 shows the left ventricle of a normal volunteer.

^{23}Na Imaging

The imaging of nuclei other than protons may provide additional diagnostic information. In particular a high concentration of ^{23}Na in the tissue is an indicator for some pathological changes. Due to the very small concentration of ^{23}Na in the body,

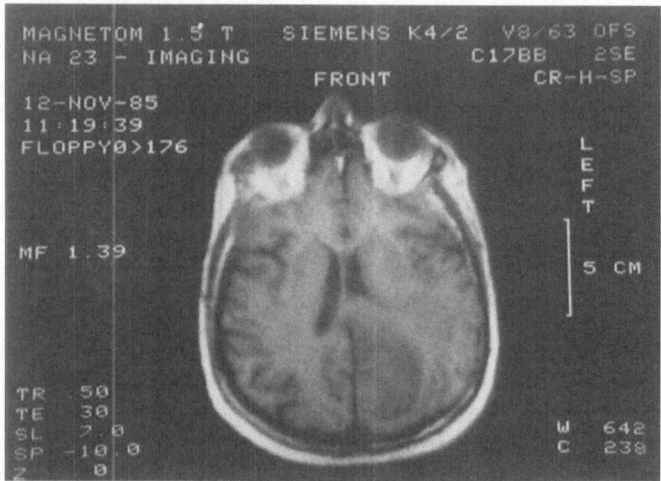

Fig. 3.26.   Proton image of a cystic brain tumour.

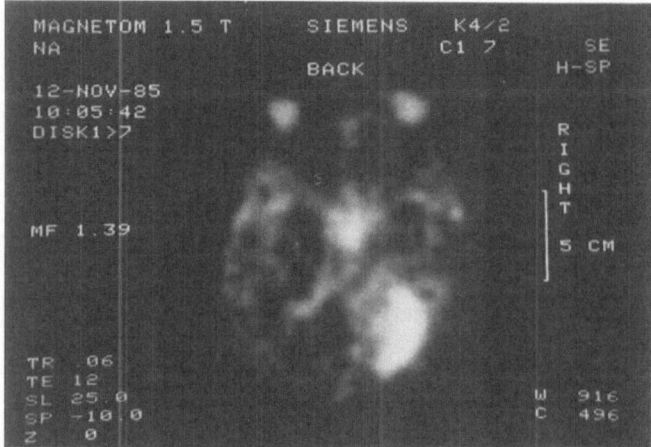

Fig. 3.27.   Sodium image of the lesion shown in Fig. 3.26.

the sensitivity for this nucleus is at least three orders of magnitude smaller than that for protons. The low signal to noise ratio (S/N) is partly improved by the choice of a larger voxel size and a large number of averages. The relaxation times T_1 and T_2 for extracellular sodium range between 10 ms and 60 ms. The intracellular component also has a very short component of about 1–5 ms. Consequently, very short spin echo delays and repetition times are needed in order to obtain good image quality within a reasonable time. The proton image and the corresponding sodium image of a cystic tumour are shown in Figs. 3.26 and 3.27 respectively. As is evident, the tumour can be localised in both images. Sodium imaging has been proven to be a very sensitive indicator for brain infarctions because of the large enrichment of this nucleus in the infarcted region.

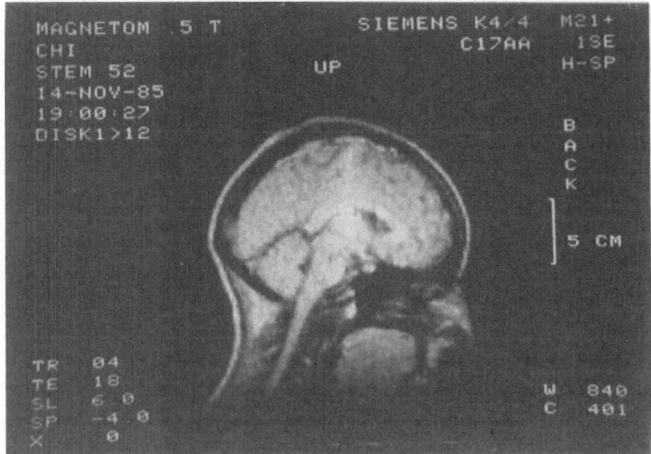

Fig. 3.28. Fast imaging with an acquisition time of 5 s — head image.

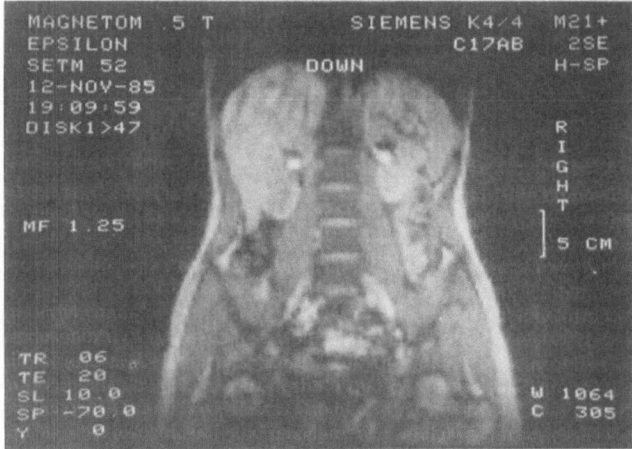

Fig. 3.29. Fast imaging with an acquisition time of 15 s — body image.

Fast Imaging

Recently developed fast imaging methods (Frahm et al. 1985; Oppelt et al. 1986) can provide imaging times of the order of 5 s, depending on the matrix size and signal averaging used. The images in Figs. 3.28 and 3.29 were obtained with a gradient echo pulse sequence combining short echo time, fast repetition and low angle excitation pulses. Dynamic processes, e.g. heart motion or uptake of contrast agents in tumours, can be investigated by these techniques.

Acknowledgements. The authors would like to thank Professor Huk from the University Hospital in Erlangen, Dr Weikl from the Cardiological Department of the University of Erlangen, Professor Zeitler from the University Hospital of Nürnberg, Professor Gerstenbrand from the University Hospital in Innsbruck and Professor Frommhold from the University Hospital in Tübingen for assistance with all of the clinical cases. Software support was also received from Professor Bloss of the Institut für Physikalische Electronik, at the University of Stuttgart. We are grateful to Professor Heohne from the Institute of Mathematics and Computer Science in Medicine, University Hospital of Hamburg-Eppendorf, for the three-dimensional computer graphics reconstruction of the heart.

References

Dixon T (1984) Simple proton spectroscopic imaging. Radiology 153: 189–194

Frahm J, Haase A, Matthaei D, Hanicke W, Merboldt K-D (1985) FLASH MR imaging: from images to movies. Radiology 157 (P): 156

Haase A, Frahm J, Haenicke W, Matthaei D (1985) 1H NMR chemical shift selective imaging (CHESS). Phys Med Biol 30: 341–344

Margosian P, Abarth J, Faul D (1984) Quick measurement of magnetic field variations within the body. Radiology 153 (P): 303

Mueller E, Deimling M, Reinhardt ER (1986) Quantification of pulsatile flow in NMR by an analysis of T_2-changes in ECG-gated multiecho experiments. Submitted to Magnetic Resonance in Medicine

Oppelt A, Grauman R, Barfuss H, Fischer H, Hartle W, Schajor W (1986) FISP, eine neue schnelle Pulssequenz fuer die Kernspin–tomographie. Elektromedica 1/86: 15

Rosen BR, Wedeen VJ, Brady TJ (1984) Selective saturation NMR-Imaging. J Comp Tomogr 8: 813–818

Stejskal EO, Tanner JE (1965) Spin diffusion measurements: Spin echoes in the presence of a time dependent field gradient. J Chem Phys 42: 288–292

Zhernovoi AI, Latyshev GD (1965) Nuclear magnetic resonance in flowing liquids. Consultants Bureau, New York

4. NMR Spectroscopy: Application to Metabolic Research

A.N. Stevens

Introduction

The aim of this chapter is to demonstrate the considerable clinical potential of NMR spectroscopy. In reviewing the course of imaging and spectroscopy over the past decade one sees that imaging has moved from visualisation of the internal structure of various fruits to high definition neurological and body imaging, where pixel sizes ($\times 0.3$ mm^2) approach those obtainable in high resolution CT. However, the clinical superiority of NMR imaging in identifying pathological information is controversial: conversely, the interpretation of NMR spectroscopy, especially ^{31}P, is well understood. Few in vivo clinical studies have been performed, principally due to the low availability of large magnets of suitable field strength and the formidable difficulties of defining precise volumes within a patient from which spectra may be taken. In reviewing the field of NMR spectroscopy it will be shown that these obstacles can be overcome and that much of the groundwork on animal studies may be readily transferred to the clinical environment. Many clinical spectroscopy systems have come into operation over the past year and a large increase in human studies is expected.

In this review the major organ/tissue systems examined using NMR spectroscopy are discussed. Many of these studies have been carried out on animals, but the extrapolation to in vivo clinical studies is clearly apparent. First, however, it is profitable to review briefly how we obtain an NMR spectrum from an area of interest.

Fundamentals of NMR Spectroscopy

Nuclear magnetic resonance (NMR) describes the property of certain nuclei which, when placed in a magnetic field, respond to the excitation of their spin states by photons of suitable frequency. Spin is a quantised property with values of $\pm 1/2$, ± 1,

± 3/2 or ± 5/2. Nuclei with odd spin values are suitable for NMR spectroscopy. NMR is more specific than other forms of spectroscopy as it involves nuclear and not electron excitation. Electrons only display measurable differences when their environment changes so the spectral information is limited. Each nucleus, however, has unique properties, one of which, the magnetogyric ratio (γ), is important in NMR.

Within a given external magnetic field (B_0) the frequency at which a nucleus resonates is dependent on the magnetogyric ratio; for example, at 1.5 Tesla (T) ^{1}H resonates at a frequency of 64 MHz, ^{19}F at 60 MHz; ^{31}P at 26 MHz and ^{13}C at 16 MHz. These large variations in frequency between different nuclei preclude simple simultaneous monitoring of such nuclei. Furthermore, there is a direct relationship between the frequency of excitation (v_0) and the magnetic field experienced by the nucleus ($v_0 = \gamma/2\pi \, B_0$). If the external magnetic field increases, the frequency of excitation (the Larmor frequency) is increased. At a given applied field B_0 a nucleus will experience a range of different magnetic fields, B_{eff}. These are brought about by the surrounding atoms whose electrons may either enhance or oppose the applied field B_0. These "shielding" and "deshielding" effects are unique for each molecule and define the chemical shift range, the frequency range over which a given nucleus will resonate. This may vary from 500 Hz for protons at 1.5 T to 10 KHz for ^{19}F at the same field. Within this "waveband" the same nucleus in different environments will be excited at different frequency and give rise to discrete resonances. It is not always possible to resolve these but by increasing the external field minor differences in frequency are magnified, since the Larmor frequency is proportional to magnetic field. This effect is often referred to as improving the chemical shift dispersion. In order to compare results from spectrometers operating at different fields, the resonance frequency of each peak (vr) is expressed in terms of its chemical shift in ppm from the resonance frequency of a standard peak (vs).

$$\text{Chemical shift} = \frac{(vr - vs) \times 10^6}{v_s} \tag{1}$$

The chemical shift of the standard should not change with changes in pH, or ionic strength. It is always preferable (Iles et al. 1982) that the standard should be internal, that is in the same compartment as the compounds of interest, although this is not always possible. Commonly phosphocreatine PCr is the reference compound for ^{31}P studies and TMS (trimethylsilane) is used for ^{1}H and ^{13}C studies.

When placed in a magnetic field, B_0, all nuclei align either parallel or antiparallel to the field. Possessing spin, the nuclei precess on their axes like tops at their resonant frequencies. A radio-frequency aerial, the coil, is able to both transmit and receive radio-frequency radiation (RF). The coil is placed in such a manner that a short burst of high energy RF in the xy plane leads to a local magnetic field along the x-axis. Spins previously aligned to the main field along the z-axis now align on the x-axis. The energy required to bring this about is referred to as the 90° pulse and is a function of both the RF power and length of pulse transmission. If this coil or another now acts as a receiver, spinning nuclei in the xy plane will induce a current in the coil, the frequency of which is dependent on the frequency of precession and is unique for each molecule. Sampling of this current over time followed by Fourier transformation gives us the NMR spectrum.

The signal amplitude of a particular NMR resonance is proportional to the number of nuclei giving rise to the resonance. Results are commonly reported as the ratio of two or more resonances or alternatively in absolute terms where quantitation is achieved by comparison with the peak obtained from a reference material. Commonly in ^{31}P spectroscopy the ß-ATP resonance is ascribed a value from chemical determination of the sample. There are a number of assumptions (Iles et al. 1982) inherent in this approach, and absolute quantitation is difficult to achieve.

All nuclei are subject to the process of relaxation, the parameters T_1 and T_2 which provide much of the contrast in NMR imaging.

T_1 is an enthalpy process and describes the time constant for the transfer of energy from the nuclei to the surroundings or lattice. This may be viewed as the realignment of magnetisation with the main B_0 field. In spectroscopy, like NMR imaging, the choice of pulse repetition time (Tr) can lead to saturation and diminution of signal intensity. This has important implications as the nuclei in different molecular environments have different T_1's and therefore in a multi-resonance spectrum where the relative peak intensities are important, saturation may lead to incorrect interpretation.

T_2 is an entropy process and is the time constant describing the de-phasing of spins in the xy plane from their initial position of alignment along the x-axis. The process commonly occurs through interaction with neighbouring nuclei or through micro inhomogeneity of the magnetic field T_2^*. Both lead to phase memory loss which in the case of the latter may be reversed using a spin echo sequence. The spin echo sequence was designed as a spectral editing technique to highlight or remove components on the basis of their T_2. However T_2 does not have the same importance in spectroscopy as in imaging sequences.

In general, the appropriate field strength for spectroscopy is as high as possible, at least in terms of signal to noise, as effective signal to noise increases approximately linearly with field. Increasing the frequency may in many cases also improve the spectral dispersion.

A variety of successful clinical spectroscopic investigations have been carried out at fields of ≥ 1.5 T. Indeed, given the problems of siting large bore systems with fields in excess of 2.0 T and the requirement for proton image information to assist spectroscopic localisation it is probable that most clinical studies will be carried out between 1.5 T and 2.0 T.

Localisation of the Spectroscopic Information

It would be desirable to use a simple imaging technique in which a frequency selective read gradient is applied to gain chemical shift and therefore spectroscopic information from a given location. Unfortunately the sensitivity of many of the nuclei preclude this approach (Table 4.1), sodium being the exception. Maudsley et al. (1984) calculated that an imaging time of 1 h would be required to obtain a phosphorous image with spatial resolution of 2 cm while for sodium 8 mm resolution may be obtained in half the time. Even proton chemical shift imaging is only suitable for detection of water and lipid resonances.

If the coil can surround the tissue of interest there is little problem in obtaining spectroscopic information. For example, coils similar to those used as "head coils" in NMR imaging have been used to study the heart and kidney but this requires surgery to enable placement of the coil and furthermore only gives a spectroscopic signal of the whole organ.

Table 4.1. Properties of biologically important nuclei

Nuclei	Spin quantum number	Relative sensitivity for equal number of nuclei	Natural abundance	Frequency (MHz) at 1.5 T	Tissue metabolite concentration
^{1}H	½	1	100	63.86	< 2 mM, lactate
^{13}C	½	1.6×10^{-2}	1	16.06	< 1 mM, sugar phosphates
^{19}F	½	0.83	100	60.08	< 1 mM, labelled drugs
^{23}Na	3/2	9.3×10^{-2}	100	16.89	20 mM tissue 140 mM extra-cellular
^{31}P	½	6.6×10^{-2}	100	25.85	< 25 mM PCR, < 5 mM ATP

To overcome this limitation the Oxford group introduced surface coils (Ackermann et al. 1980). This was an important development as it allowed studies of NMR to be carried out on whole animals and humans non-invasively. Their definition of a surface coil was a flat 1 – 3 turn loop of wire which was placed on the area under study. However, as in imaging, the term surface coil has come to be associated with any coil closely coupled to the region of interest and recently even resonator structures derived from electron spin resonance (ESR) have been used to permit localisation (Grist and Hyde 1985).

The surface coil produces a very localised sensitive volume, allowing selected regions of an organ to be studied. The signal to noise ratio is much better than with conventional coils as noise only derives from the volume giving rise to the signal, therefore maximising sensitivity. The success of the method is displayed by the widespread use to which it has been placed, with the most accessible organs' systems having been examined. The limitation of the surface coil approach is that it generates an RF field which varies in power through the surrounding space (commonly observed in NMR imaging as a decline in signal intensity away from the coil). This provides a serious problem when studying deep-seated tissues in situ as there are undesirable contributions to the spectra from intervening surface tissue. Recently, several remedies have been proposed to improve surface coil localisation.

The first was TMR, topical magnetic resonance (Gordon et al. 1980). TMR is a simple sensitive point method that uses a combination of static non-linear magnetic field gradients to profile and restrict the effective homogeneous volume of the applied field B_0. If this sensitive volume is located in a sample then the RF coil will detect signal from the whole of the sample but the signal will consist of two components, narrow lines from the compartment of interest and inhomogeneously broadened lines from the rest of the sample. These broad lines may be removed by the computer (convolution differencing). One constraint on the method is that the inhomogeneity of B_0 throughout the sensitive volume must be less than the smallest difference in chemical shift that is anticipated.

TMR was rather crude and recently more sophisticated techniques making use of NMR imaging experience have come to the fore.

In the DRESS technique (Bottomley et al. 1983) a large RF coil is used to provide a uniform RF excitation field over the sensitive volume and a smaller diameter surface coil is used for detection. Depth resolution is achieved by incorporating at least one selective excitation pulse in the RF pulse sequence. The selective excitation pulse is applied during the presence of a magnetic field gradient pulse to excite a flat plane of nuclei parallel to the plane of the surface coil. The extent of the sensitive volume within the selective plane is determined only by the diameter of the surface detection coil and the distance of the coil from the sensitive plane. Within the repetition time of excitation, typically up to 8 s, a number of locations may be sampled enabling multi-slice DRESS (SLIT DRESS) (Bottomley et al. 1985).

By combining selective RF pulses with pulse field gradients along the three cartesian axes both VSE (Aue et al. 1984) and ISIS (Ordige et al. 1986) allow a volume of interest to be selected within the body. Moreover, the position of the volume may be varied by altering the frequency of RF excitation. In VSE selective and non-selective 90° pulses are used to excite a volume of interest whilst ISIS uses selective 180° pulses so that signals from undesirable locations are cancelled over an eight set sequence. The requirement for a broad-banded pulse and associated RF power will probably restrict the use of VSE in clinical studies.

The use of pulsed field gradients can lead to impairment of spectral quality due to mixing of the spatial and chemical shift information and, additionally, resonances with short T_2 times may be lost.

An alternative approach is to make use of the inhomogeneous RF profile of the surface coil.

Rotating frame zeugmatography uses two concentric surface coils; the larger one to transmit, the smaller to receive (Styles et al. 1985). A series of data sets are obtained as a function of the pulse angle which on Fourier transformation yield a set of spectra from a series of slices parallel to the plane of the surface coil. This technique has the advantage of requiring no gradient coils but has problems with edge definition, and it is difficult to correlate the spectral slice with an image.

Bendall has used a number of schemes to obtain spatial resolution using a surface coil and combinations of the standard methods of inversion recovery and spin echo in which the refocussing 180° pulse is recycled through all four phases ($\pm x \pm y$) (Bendall and Gordon 1983). This effectively discriminates heavily against sample regions where the pulse angle departs from 180°. The result is a shaped pulse where, for example, signal is only acquired from regions experiencing a pulse angle of 90° ± 10°. However, to date only phantom studies have been carried out and its application to clinical situations has not been demonstrated.

These techniques, while not a complete solution, do open up the prospect of the in vivo studies discussed in this review developing into practical methods of diagnosis of human disease.

Clinically Useful NMR Nuclei

The nuclei that are observable by NMR and are clinically useful are numerous; however, those likely to be important are probably limited to ^{1}H, ^{13}C, ^{19}F, ^{23}Na and ^{31}P. We may divide these into two groups: those which are naturally abundant in the body (^{1}H, ^{13}C, ^{31}P, ^{23}Na), and those which need to be administered (^{19}F). It is in fact more logical to put ^{13}C into the second category as in most cases it is administered as a tracer.

Most work to date has involved the ^{31}P nucleus. This nucleus is 100% naturally abundant and has a relatively high NMR sensitivity although only 7% that of protons. Typical spectra shown in Fig. 4.1 allow observation of a selection of important high energy phosphates involved in the transduction of energy within the cell. Quantitation of these resonances therefore allows us to study the major biochemical pathways leading to their formation, namely oxidative and anaerobic respiration. Clinically, ischaemia/hypoxia is the most important modulator of these processes but many forms of work, for example muscle exercise and chemical synthesis, also alter their concentration. In addition, inborn errors of metabolism may be identified. Secondly, lipid metabolism may be examined by observing the broad underlying component that arises primarily from phospholipids, and the PDE (phosphodiester) and PME (phosphomonoester) resonances. Thirdly, exogenous compounds, for example sugars, may often be directly observed in the PME region as their phosphate derivatives.

Additionally, NMR spectroscopy provides an alternative method to micro-electrodes or dye binding studies for the measurement of intracellular pH, and is probably the most suitable method for clinical studies. If a group containing an ionisable proton is in an environment whose pH is close to the pK of that group, then the chemical shift of signals from nuclei in the group may change with variations in the pH of the environment.

Pi has a pH of 6.8 and thus exists at physiological pH as the ions HPO_4^{2-} and $H_2PO_4^-$; the signal from HPO_4^{2-} is shifted downfield by 2.25 ppm from that of the $H_2PO_4^-$. However, the two forms are in rapid exchange so that a single Pi peak is formed with a chemical shift reflecting the relative proportions of the two ions. Thus a titration curve of the relationship between the chemical shift and pH can be obtained and the nucleus may be used as a pH indicator (Fig. 4.2). Since all cells contain Pi, this method has many potential applications in the study of intracellular pH (pHi). In some circumstances it is possible to derive similar data from organophosphate compounds with suitable pK values; they also give single peaks whose chemical shifts reflect the pH of the environment, for example 2-deoxyglucose-6-phosphate.

This method of pH measurement has recently been further validated (Williams et al. 1985). Studies using doubly tuned coils for ^{31}P and ^1H have allowed comparison of pH values obtained from Pi to those from the upfield ^1H resonances of anserine (+7, +8 ppm) which derives from a group titratable in the physiological range. The normoxic pH in rat leg muscle obtained by ^{31}P was 7.06 ± 0.03, in good agreement with that measured by ^1H: 6.98 ± 0.04 and 7.04 ± 0.04 (n = 13). Similar validations have been achieved using ^{19}F (Civan et al. 1984).

Fig. 4.1. ^{31}P spectrum of rat **a** hind limb muscle, **b** liver and **c** implanted tumour line, all obtained using a surface coil. The spectra were recorded at 32 MHz (1.9 T) and represent the accumulation of 240 transients at 2-s intervals. Peak *A* contains signals from a number of phosphate monoesters (PMEs), predominantly phosphorylcholine but commonly glucose-6-phosphate, glycerol-3-phosphate and AMP. Peak *B* is from Pi; peak *C* corresponds to glycerophosphoryl-ethanolamine (GPE) and to glycerophosphorylcholine (GPC). Peak *D* not observed in liver is due to phosphocreatine (PCr). Peak *E* is from the γ-phosphorus of ATP and also from the ß-phosphorus nuclei of ADP, whereas peak *G* arises from the ß-phosphorus of ATP. Thus, by subtracting the integral of peak *G* from that of peak *E*, a value for ADP can be obtained. Peak *F* is from the α-phosphorus of ATP and ADP and the two adjacent phosphorus nuclei of NAD and NADP. It is also likely that we observe GTP, which in liver, for example, is present at 10% of the level of ATP and resonates at a similar frequency to ATP. It is evident that these signals in liver and tumour are superimposed on a broad component, unlike the flat baseline in spectra from muscle. This has been ascribed to the slow tumbling phospholipids.

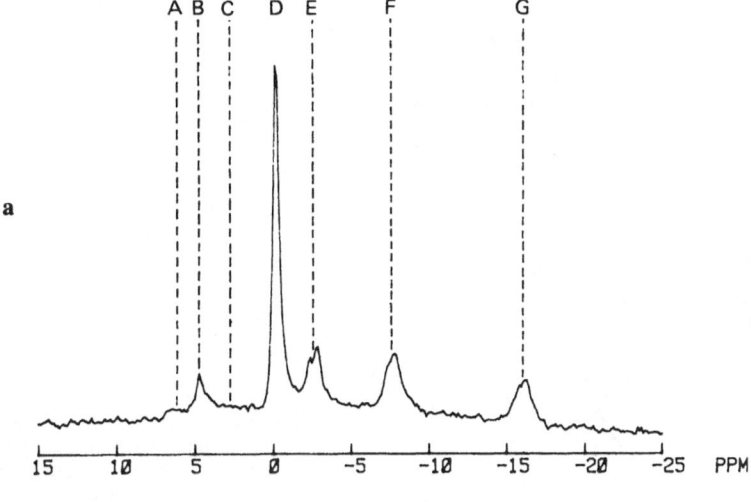

a

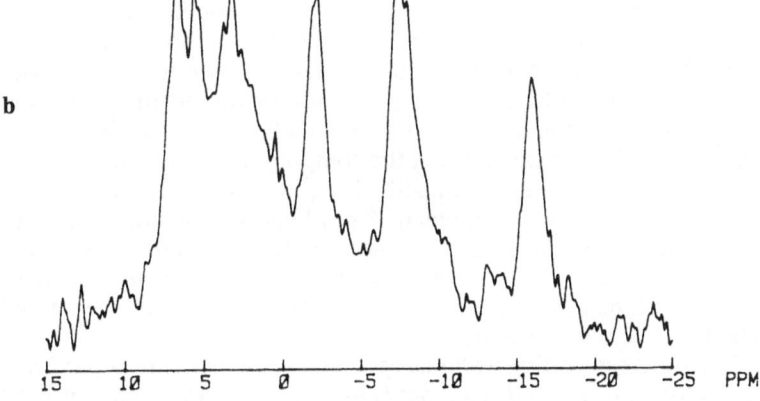

b

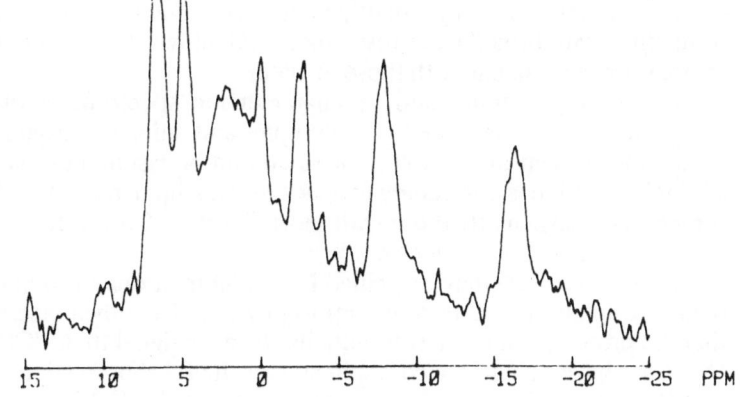

c

Fig. 4.1.

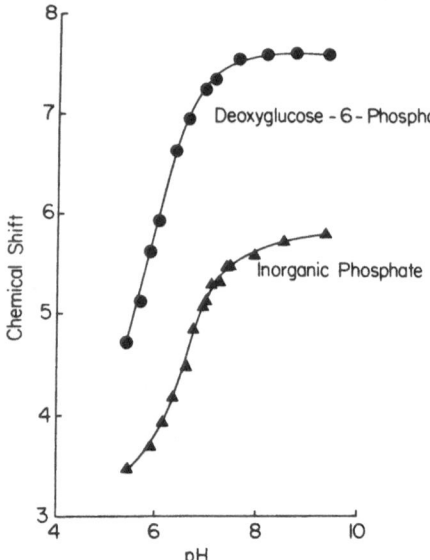

Fig. 4.2. Variation of the chemical shift of inorganic phosphate and deoxyglucose-6-phosphate with solution pH at 37°C. The medium contained 150 mM K+, 5 mM Na+ and 1 mM Mg2+. The chemical shift is expressed relative to phosphocreatine.

The study of in vivo proton NMR spectroscopy has recently received increased interest. The proton nucleus is the most sensitive of those found in biological systems, and suitable spectra may be acquired in less than 1 min. The chemical shift range is, however, small and the spectrum is complex due to the large number of proton-containing compounds in the cell, many of which give rise to broad overlapping peaks. A further problem is that the compounds of interest are present at millimolar concentrations, and are swamped by the intense signals from the protons in water (100 molar) and lipid (1 molar). It has been in this area that important practical developments have been made — noticeably the advent of 16-bit analogue to digital convertors which, when associated with various water suppression techniques (1331, selective irradiation and spin echo), allow us to visualise metabolites present at relatively low concentration.

The most interesting water suppression sequences are those involving composite pulses, notably the 1331 sequence (Hore 1983). By manipulating the pulse angles and adjusting their spacing to correspond to the offset of the region of interest from the water resonance it is possible to attain a suppression of the H_2O resonance by a factor of 10 000. A spin echo can also be incorporated to remove unwanted signals from compounds with short T_2, notably CH_2. The advantage of the method is that it is compatible with broad water resonances and allows detection of compounds whose protons are in exchange with those of water.

Once water is eliminated, a characteristic spectrum is obtained (Fig. 4.3) comprising of aromatic +4 − +8 ppm and aliphatic regions. In theory any compound present above 100 μ molar should be visible but many of the "spectral editing" techniques for removal of water and lipid may be selective for certain compounds, particularly those with short T_2 times. Currently, the limit of detection is that of the phosphorus nuclei.

Further selectivity may be gained by the indirect observation of ^{13}C tracers from their J coupling (i.e. C-H coupling) to protons. The advantage of this technique is that the proton produces a substantially stronger signal than [13]C. Protons bonded to ^{13}C atoms have a signal which is split as a result of the interaction of neighbouring ^{13}C and ^{1}H spins. [13]C decoupling, i.e. irradiation at the ^{13}C frequency, removes these

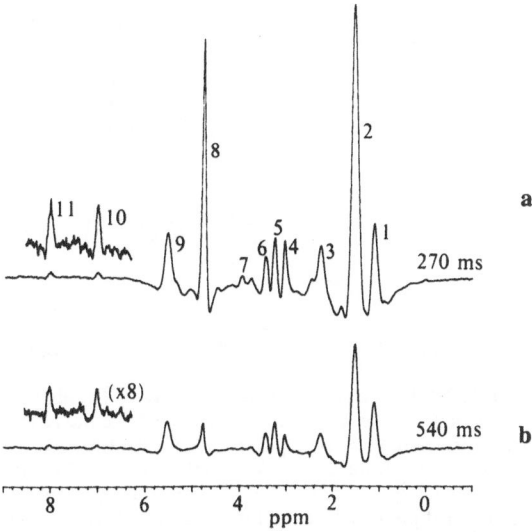

Fig. 4.3. ¹H spectrum of rat leg muscle in vivo. The spectrum was recorded at 360 MHz (8.5 T), and is the result of 16 acquisitions taking 30 s. The water resonance 8 has been greatly reduced by the use of a spin echo sequence with a delay (equivalent to Te) of 270 ms (**a**) and 540 ms (**b**). Signals were assigned as follows: *1, 2, 3* and *9* various fat protons; *4* and *7*, the N-CH$_3$ and N-CH$_2$ protons of phosphocreatine and creatine; *5* and *6*, predominantly taurine; *8*, water; *10* and *11*, anserine. The resonances are given relative to TMS (Williams et al. 1985).

splittings. Subtraction of undecoupled and decoupled spectra enables the ¹H resonances adjacent to ¹³C atoms to be highlighted.

Sodium (²³Na) with a natural abundance of 100% can be measured with higher sensitivity than ³¹P. As a hydrated ion it is seen as a single resonance. Workers have shown that the intra- and extracellular space may be identified by differences in T$_2$ relaxation times, caused by binding of the intracellular sodium. Alternatively, the extracellular space may be probed using T$_1$ relaxation agents, notably dysprosium salts, allowing differentiation of the two compartments on the basis of T$_1$ (Gupta and Gupta 1982). The high sensitivity of sodium and its short T$_1$ and T$_2$ behaviour make it a suitable nucleus to image (Hilal et al. 1983). An interesting result showing increased intracellular concentration of sodium in metastatic tissue has been noted; however, no concentration values have been given. The increases have been ascribed to increased mitotic activity but this explanation is certainly not universally supported.

Further information may be gained by using ¹⁹F cation probes based on the chelating agent BAPTA (Smith et al. 1983). These compounds are susceptible to shift on cation binding and allow ionic concentrations to be determined. They are rendered selectively intracellular by incorporation of an ester group within the molecule which is cleaved on entry into the cell, locking in the molecule. Intracellular Na levels of 15 mM have been determined in preparations of lymphocytes.

The sensitivity of ¹³C is less than 2% of protons and is compounded by the low natural abundance of the isotope (1%). Typical natural abundance spectra therefore derive solely from compounds present at high concentration, primarily the carbon backbone of proteins and lipids, although glycogen has also been detected (Fig. 4.4). A second problem concerns the potential coupling or interaction of the ¹³C nucleus to neighbouring proton and carbon nuclei leading to splitting of the resonance. This

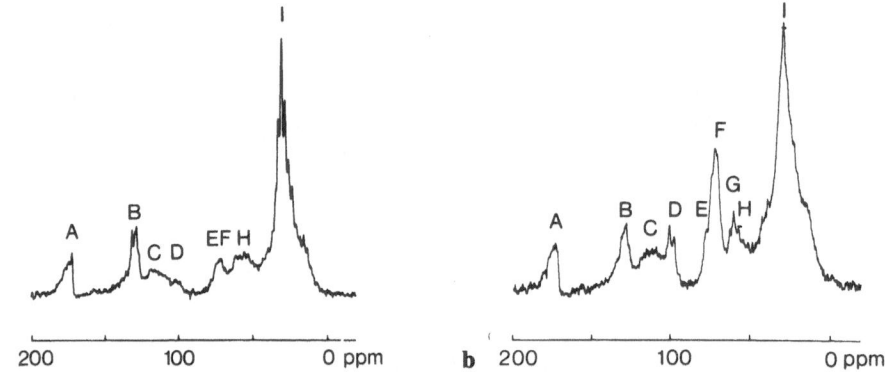

Fig. 4.4. **a** Proton decoupled 50.24 MHz (4.7 T) ^{13}C NMR spectrum of fed rat liver in vivo. The aliphatic region (0–100 ppm) contains a major peak at 30 ppm (*I*) which almost entirely derives from the $(CH_2)_n$ repeating unit of fatty acid chains situated in triglycerides and phospholipids with some contributions for similar units in amino acids. Resonance *H* at 54 ppm derives from choline. Within the region 50–110 ppm many of the substrates and products of intermediary metabolism are visible, notably the sugar phosphates. Commonly the C-1 carbon of glycogen is observed at 101 ppm (*D*). Left of this region we observe carbon atoms in delocalised electron environments, notably the γ and δ carbons of histidine, phenylalanine and tryptophan (*C*), the double bonded carbons primarily from phospholipids (*B*) and carbonyl carbons of proteins (*A*). The resonances are given relative to TMS. **b** Fed rat liver from an animal with glycogen storage disease. Notice the peaks *D, E, F* and *G* deriving from glycogen.

may be eliminated by low level RF irradiation at the coupled frequency (decoupling); in human studies, however, this may lead to excessive RF deposition.

The sensitivity of ^{13}C may be increased using enriched compounds and such tracer experiments may be carried out to observe the fate of these compounds or the metabolic capacity of an organ system to deal with them. Dr. R. Shulman, Cohen and collaborators have been pre-eminent in such studies. As described earlier, one may also use the coupling of the carbon nucleus to its neighbouring proton to observe tracer ^{13}C compounds using the sensitivity of proton MR – so-called cross polarisation (Rothman et al. 1985).

Fluorine (^{19}F) is not present in naturally occurring biochemicals and therefore must be observed after administration of exogenous compounds. This gives a selective advantage over other nuclei in that there is no background to contend with. Furthermore, the fluorine nucleus is almost as sensitive (83%) as the proton nucleus and is 100% naturally abundant. It also has an exceedingly large chemical shift range which means that minor alterations in the environment of the nucleus are clearly shown by the presence of shifted resonances as shown in Fig. 4.5. Studies to date have concentrated on antimetabolites (2-*F*-deoxyglucose) and various drugs including anaesthetics (halothane) and anticancer agents (5FU). Many other classes of drugs, for example the analgesics, neuroleptics and antibiotics, are well represented with fluorine derivatives.

In Vivo Studies

Much of our understanding of in vivo processes derives from studies carried out in vitro. The interest in NMR spectroscopy arises from the desire to follow metabolism in vivo to more fully understand changes in physiological and pathological states.

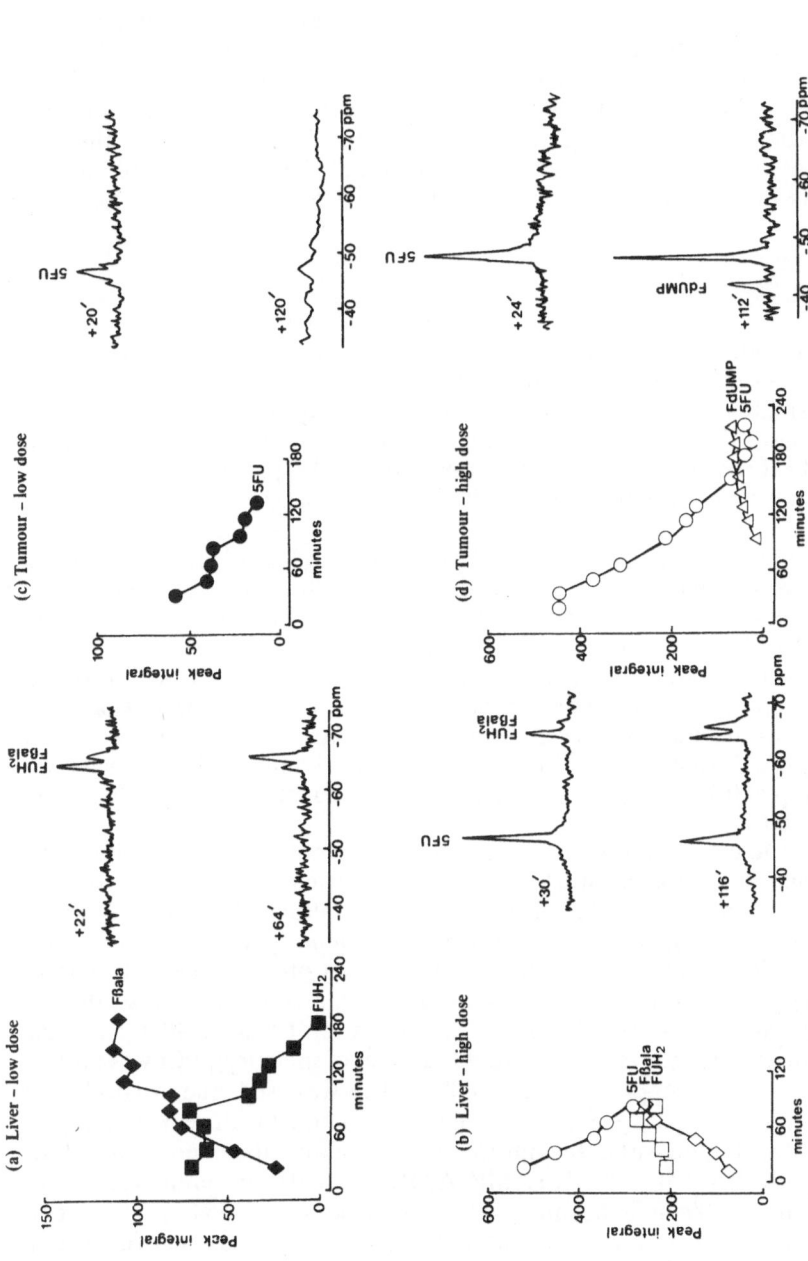

Fig. 4.5. Uptake of 5-fluorouracil (5FU) into liver and implants of Lewis lung tumour in C57 mice monitored by ^{19}F NMR spectroscopy after the IV injection of 30 mg kg^{-1} 5FU (**a, c**) and 180 mg kg^{-1} 5FU (**b, d**) into the jugular vein. Times refer to time after injection of 5FU. FdUMP ($\triangle$), the major anabolic product, arises primarily from the conversion of 5FU (O,●). The resonance probably contains contributions from the corresponding di- and tri-phosphates. Dihydrofluorouracil (FUH$_2$) ($\square$ and ■) represents the first ring cleavage product of the 5FU catabolic pathway. Fluoro-ß-alanine (F-ß-ala) ($\lozenge$,♦) is the end product of 5FU catabolism. Chemical shift values have been derived from the assignment of 5FU and F-ß-ala, which provide non-titratable internal references. All shift values are given relative to a pure solution of fluorotryptophan (Stevens et al. 1984).

Muscle

A major application of spectroscopy has been in the study of the bioenergetics of animal and human muscle. It is preferable to the technique of needle biopsy sampling for chemical estimation as this is painful and, in many cases, not feasible, e.g. when following a patient's response to exercise. The larger sample volume provided by spectroscopic study is also advantageous in that more representative sampling is possible. This could be particularly important in the case of myopathic muscle in which the tissue is heterogeneous on a microscopic scale. Furthermore, the time taken for biopsy sampling (5 s) leads to degradation of the metabolites of interest. Specifically, the ratio PCr/Pi is consistently higher when determined by spectroscopy.

Exercise

Chance et al. (1980) and Cresshull et al. (1981) were the first to study human muscle by ^{31}P NMR during rest and exercise. Phosphocreatine (PCr) and inorganic phosphate (Pi) ratios were measured in forearm and calf muscles in the resting state and after exercise. During exercise the breakdown of PCr was mirrored by similar increases in Pi. Little change in pH was observed as the production of protons, consistent with lactate formation, was small and was offset by a net removal by PCr hydrolysis.

After vascular occlusion, exercise caused a 66% drop in PCr/Pi after 30 s. During a further 5 min of ischaemia the pH declined from 6.98 to 6.70, suggesting that anaerobic glycolysis was taking place, leading to formation of lactate with subsequent acidification. No PCr recovery occurred, indicating the requirement of oxidative phosphorylation for synthesis. In a further study, Chance et al. (1981) looked at work output in the exercising human forearm more closely; work was by wrist flexion. PCr/Pi ratios varied from approximately 20 at rest to unity during exercise. Between low work rates, termed "comfortable exercise", and high work rates, at which cramp was felt, a small drop of 0.1 pH units was experienced, corresponding to a fall in the PCr/Pi ratio from 1 to 0.67.

Over the last few years the Oxford group have concentrated on examining the flexor digitorum superficialis muscle, at rest, during aerobic dynamic exercise, ischaemic exercise and periods of recovery (Taylor et al. 1983). Using surface coils of not more than 25 mm diameter, it was possible to obtain spectra solely from this muscle with no significant contamination of signal from other muscles. In spite of wide individual variation, there were general trends in responses. At rest, the pH was 7.03 ± 0.03 and the PCr/ATP and Pi/ATP ratios were 1.7 and 0.48. On aerobic exercise (by squeezing a bulb set at different pressures of mercury), PCr was broken down to Pi and the pH dropped when over 60% of PCr was used, due to a decline in the effective buffering capacity of the tissue. During this time the Pi peak broadened, indicating pH distribution within the muscle, and a shoulder appeared to the left of the main peak which was attributed to AMP. ATP remained stable until over 90% of the PCr was used. When ischaemic exercise was performed, no post-exercise recovery occurred until aerobic metabolism was possible. PCr recovery (indicating oxidative phosphorylation) was much quicker than pH recovery (indicating removal of H+ ions); indeed, the muscle actually became more acidic at first due to the production of H+ from PCr and ATP synthesis. There was a rapid fall in Pi which was attributed to Pi entering the mitochondria and becoming invisible to NMR.

While there were large variations in the oxidative and glycolytic capacities of different individuals it proved possible to establish the set of criteria outlined above that were characteristic of healthy muscle. This knowledge has proved important in the interpretation of data from individuals suffering from musculoskeletal disorders. Some of the clearest observations have been seen in patients suffering from metabolic and mitochondrial myopathies.

Metabolic Myopathies

Ross et al. (1981) investigated a suspected case of McArdle's syndrome, a glycogen storage disease in which muscle glycogen cannot be adequately broken down because of phosphorylase deficiency. Spectra were obtained at rest, during aerobic exercise, anaerobic exercise and recovery. Exercise was by flexing the hand. The intracellular pH of the patient was 7.2 at rest, higher than 5 controls in whom pH_i was 7.02 ± 0.01. During both aerobic and anaerobic exercise the decrease in PCr and the increase in Pi were more extensive than in the controls. During anaerobic exercise there was no measurable decrease in intracellular pH in the patient's muscle while the healthy volunteers showed a decrease from 7.2 to 6.7 ± 0.1. The results were consistent with McArdle's syndrome. Acidification was not observed during exercise because glycogen was not mobilised and so only negligible amounts of lactic acid were produced.

Similar results were observed in patients suffering from type III glycogen storage disease, another metabolic myopathy, in which the debrancher enzyme (amylo-1,6-glucosidase) is deficient (Radda 1984). Acidosis was not observed during exercise, whilst the alkalosis was smaller (pH = 0.05) than seen in McArdle's syndrome. All three patients had an elevated resting pH.

Mitochondrial Myopathies

Gadian and co-workers (1981) investigated a patient with mitochondrial myopathy. Earlier investigations on this patient had pointed to a lesion in electron transport being the primary complaint. The resting pH_i value of the patient was little different from that in controls, however, and the concentration of PCr was within control values but the Pi concentration in the muscle was elevated. This gave PCr/Pi ratios in the patient of 3.8 ± 0.7 (four successive measurements) as opposed to 10.0 ± 2.7 in the controls.

In a second study of two patients with mitochondrial myopathies, Radda et al. (1982) showed that structurally abnormal mitochondria and the inability to oxidase NAD-linked substrates in these patients were reflected in the NMR spectra obtained on exercise. In this case the concentration of PCr decreased and that of Pi increased more rapidly than in the controls. The patients took ten times as long for their PCr/Pi ratios to return to the resting level (30 min). They attributed this slow rate of recovery to depressed oxidative phosphorylation.

Similar results were obtained in studies of a selection of mitochondrial myopathies, namely slow recovery of PCr after exercise and a reduced phosphorylation state (i.e. ATP/Pi or PCr/Pi) ratio in the initial resting spectrum.

Exercise studies of human muscle have contributed to the fundamental understanding of basic bioenergetics. However, the information has been limited due to the resolution of the NMR method which makes visualisation of changes over a small time period (< 10 s) impossible. The studies of metabolic disorders have been interesting and have shown that the NMR parameters responded in the expected manner. NMR is, however, probably impractical as a day-to-day tool for diagnosis of these disorders.

Muscular Weakness and Dystrophy

Perhaps the most unexpected results have come from the examination of patients with undiagnosed muscle problems (i.e. weakness, fatiguability, pain); approximately 20% have shown some kind of deviation from the normal NMR pattern. Among these a significant group is formed by patients who exhibit abnormally rapid intracellular acidosis at the early stages of aerobic exercise (Arnold et al. 1984). In recovery, the resynthesis of PCr after exercise was somewhat slower than that observed in healthy controls following the same exercise protocol and is attributed to a mitochondrial dysfunction. This condition appears to be associated in some way with a previous viral infection which is known, in some cases, to lead to muscular weakness.

In the condition polymyositis, an inflammatory muscle disease causing the sufferer to tire easily on exercise, a high level of PCr breakdown was observed (Fig. 4.6). The data suggest an inefficiency in the use of energy during muscle contraction rather than an inability to regenerate muscle energy stores (Iles et al. 1982).

Major changes in the resting concentrations and compositions of phosphate-containing compounds have been observed in Duchenne dystrophy (Edwards et al. 1982). In dystrophic boys the levels of PCr/Pi and ATP were low with PCr being greatly reduced; pH was slightly higher than normal. The decrease in PCr could be attributed to the replacement of muscle by fat, as shown by proton spectra of muscles, or by fibrous tissue, implying that the metabolism of remaining muscle in dystrophies is changed and that the ratio of PCr/ATP is reduced in dystrophic muscle. NMR contributes to the understanding of this condition by showing that, compared to the metabolic and mitochondrial myopathies, a reduction in phosphorylation potential is not a cause of muscular dystrophy but that such an energy deficit is a reflection of the dystrophic state.

Brain

Accurate studies of human brain metabolism are only possible using either NMR or emission CT and to date it is one of the most popular systems to be studied. Initial studies on rodents (Ackerman et al. 1980) were quickly extended to ^{31}P NMR studies of both

---

Fig. 4.6. **a** ^{31}P NMR spectra (32 MHz) of the forearm flexor muscles of a polymyositis patient at rest, during exercise and during recovery. The exercise was the squeezing of a rubber bulb to a standard pressure every second: the muscles were relaxed during data acquisition. Spectrum a) (*bottom left*) was obtained over a 4-min acquisition period (i.e. 240 1-s pulses), spectrum b) during 60 1-s contractions (following 30 s relaxation). Spectra d), e) and f) were obtained after increasing recovery periods, again with 1-s pulses, following a further 1-min exercise period [spectrum c)]. In each case the number above the phosphate peak represents pH$_i$ from the chemical shifts of that peak. **b** Spectra obtained in the same manner as in (**a**) except that the exercise period was longer (Iles et al. 1982).

Fig. 4.6.

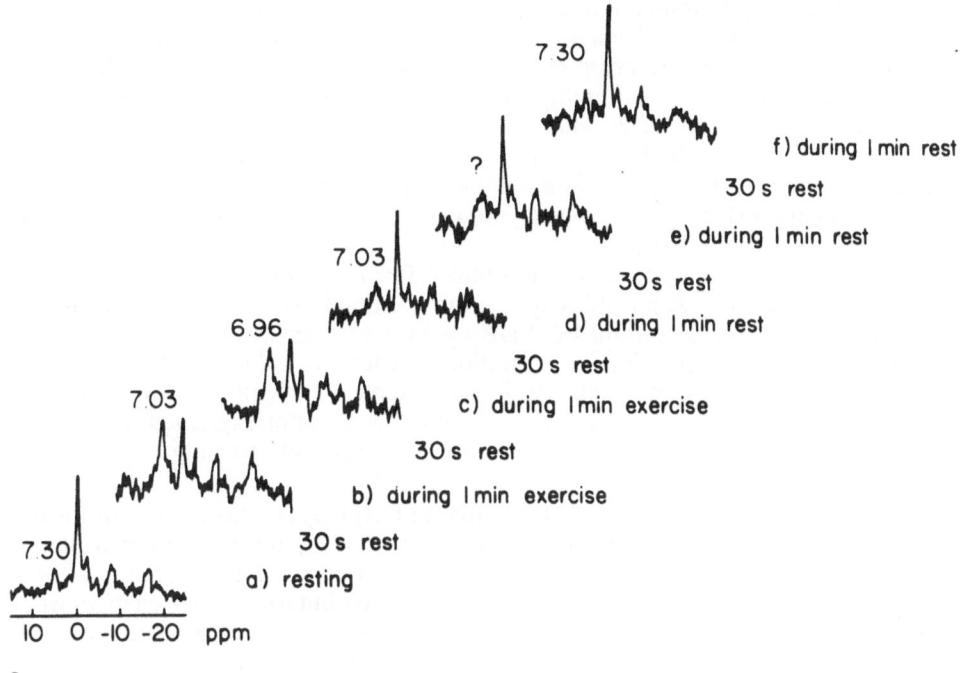

7.30

?

f) during I min rest

30 s rest

e) during I min rest

7.03

30 s rest

6.96

d) during I min rest

7.03

30 s rest

c) during I min exercise

30 s rest

b) during I min exercise

7.30

30 s rest

a) resting

10 0 -10 -20 ppm

a

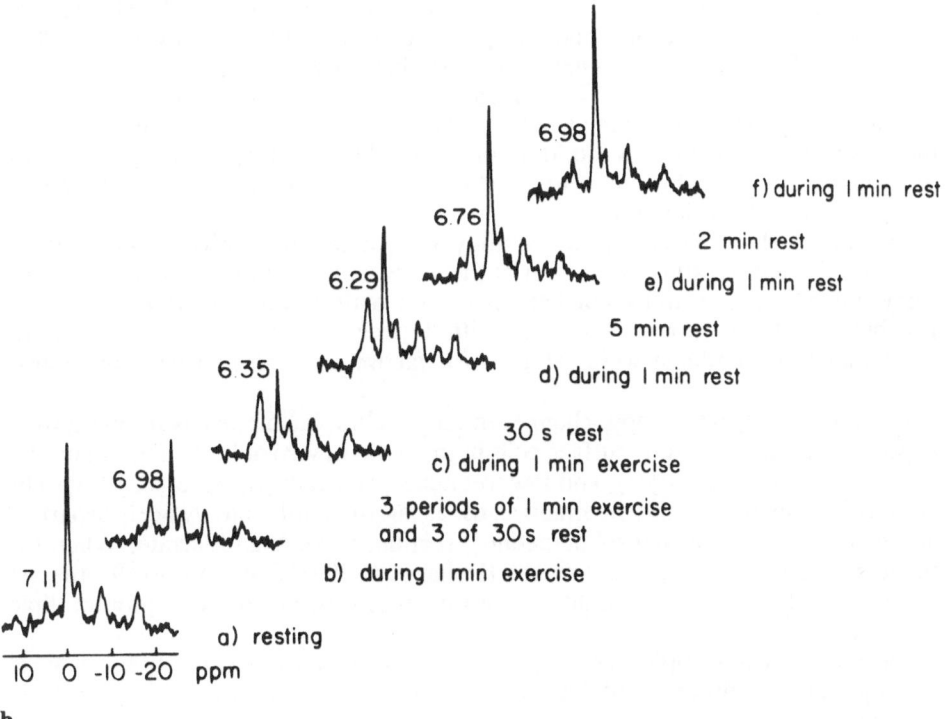

6.98

6.76

f) during I min rest

6.29

2 min rest

e) during I min rest

6.35

5 min rest

d) during I min rest

30 s rest

c) during I min exercise

6.98

3 periods of I min exercise
and 3 of 30 s rest

7 11

b) during I min exercise

a) resting

10 0 -10 -20 ppm

b

neonatal human brain (Cady et al. 1983) and more recently, with the advent of wider bore machines, adult brain (Bottomley et al. 1983; Radda 1984).

Surface coil studies have been used to examine a number of models of hypoxia and ischaemia. In ischaemia (Naruse et al. 1984), rapid loss of high-energy phosphate levels was observed with a flattening out of the simultaneously monitored EEG. The ischaemic period lasted for 30 min. On restoration of flow by untying the carotid arteries, the levels of high-energy phosphates returned rapidly to near normal. However, the EEG remained abnormal, even after 12 h. The authors gave no clear explanation of the discrepancy but suggested that brain death cannot be solely judged by the flattening of the EEG.

One interesting finding in the brain is that ATP and PCr may fall in concert after the onset of ischaemia. A possible explanation was that the enzyme, creatine kinase, mediating the phosphorylation of ADP by PCr was of lower activity in brain. Shoubridge et al. (1982), using the saturation transfer technique, and Balaban et al. (1983), using 2D NMR, were able to follow chemical exchange and disprove this suggestion. The simultaneous fall in ATP and PCr is better explained in terms of tissue heterogeneity where local energy demands are able to fully deplete PCr, leading to a fall in ATP.

Prichard et al. (1983) investigated the effects of hypoglycaemia induced by insulin administration, hypoxia and status epilepticus caused by intravenous bicuculline. During hypoglycaemia and hypoxia a reversible decrease in phosphocreatine stores accompanied by an increase in Pi was detected. Similar but irreversible changes were observed in status epilepticus. ATP levels fell only markedly in hypoglycaemia and this was accompanied by an increase in intracellular pH. While changes were reversible the preparation did not return to its control state; pH_i remained low while Pi did not completely recover. The EEG remained abnormal. During the hypoglycaemic state the phospholipid peak decreased and this was attributed to switching of fuel sources from glucose to fat during stress.

Ackerman et al. (1984) have looked at the rat brain in vivo under normal conditions and when the animal's blood had been replaced by a perfluorocarbon blood substitute. They saw no differences in the NMR spectra or the NMR determined pH, allowing them to conclude that such emulsions certainly permit adequate tissue perfusion in vivo.

Other nuclei have also been used to study brain metabolism. Wyrwicz et al. (1984) followed the uptake and distribution of the anaesthetic halothane using ^{19}F NMR. They noted selective uptake of halothane into hydrophobic environments, most notably membrane lipids, where a shifted peak was observed. Interestingly, metabolites of halothane were retained for far longer than had been previously supposed.

Using proton spectroscopy, changes in the cerebral lactate levels resulting from hypoxia were monitored by surface coils in the brain in vivo at both high frequency, 360 MHz (Behar et al. 1983), and low frequency, 80 MHz (Behar et al. 1984). The authors identified a number of amino acids: aspartate, glutamate, N-acetylaspartate and alanine, as well as phosphocreatine, creatine, GABA and lactate. When the fraction of inspired oxygen was lowered (Behar et al. 1983) from 25% to 4%, the rise in the lactate resonance could be distinguished from the background lipid resonances.

Lactate has been implicated in the development of irreversible brain damage making the measurement of lactate accumulation and clearance important in diagnosis.

A recent study of mice with histidinaemia, an inborn error of amino acid metabolism, has pointed the way to the identification of phenylketonuria in the newborn (Gadian et al. 1986).

Reynolds and co-workers have, over the last two years, carried out an extensive study of the brains of newborn infants using ^{31}P NMR (Cady et al. 1983). Specifically, they have explored the evolution of NMR changes in infants who were asphyxiated during delivery or affected by other cerebral disturbances. To achieve this they designed a transport system which allowed the infants to be warmed, monitored and, if necessary, mechanically ventilated during the study. A large coil of up to 7 cm in diameter has been used to allow signals to be collected from the whole of the hemisphere adjacent to the coil.

Mean PCr/Pi in normal infants was 1.35 ± 0.22 and PCr/ATP 1.01 ± 0.14. Both of these values were low compared to adult brain. On the other hand, PME/ATP was high with a value of 2.10 ± 0.03. The intracellular pH was 7.14 ± 0.10. Children with birth asphyxia showed near normal spectra on the first day of life. The ratio PCr/Pi fell to 0.68 ± 0.34 over the next 9 days, recovering in those who survived. The fall in the ratio of PCr/Pi was attributed to poor oxidative phosphorylation. Other infants have also been studied suffering from a number of conditions. In general, whatever the diagnosis, the PCr/Pi ratio seemed to provide prognostic information. Sixteen of the first 36 infants studied had ratios below 0.8; 8 died and 8 had signs of neurodevelopmental abnormalities. Of those with ratios exceeding 0.80, 15 appeared normal, 2 died and 3 showed abnormalities.

The advent of imaging/spectroscopy systems allowed ^{31}P spectra of the adult head to be obtained (Bottomley et al. 1983). This preliminary work was criticised (Pettegrew et al. 1983; Tofts et al. 1984) on the grounds that the spectrum was of extracranial tissue, primarily the temporalis muscle. Radda et al. (1984) have shown that alteration of pulse length from 40 μs to 200 μs gave spectra resembling either that of muscle (cf. Bottomley et al. 1983) or of brain. More recently the development of localisation techniques has permitted definitive spectra to be obtained from various cerebral locations (Fig. 4.7; Bottomley et al. 1985a).

Heart

The heart has been one of the more popular areas to study using spectroscopy, with much in vivo work being carried out using the isolated Langendorf perfusion model in both the free running (with no external electrical stimulation) and paced modes. Most workers have studied the results of ischaemic insult and more recently this has been followed in vivo either in open chest models or using depth localisation.

Grove et al. (1980) introduced a four turn solenoid coil into the thorax of the animal through an abdominal incision across the diaphragm, with the coil placed around the ventricles. Ventilation was maintained mechanically. After an initial spectrum, ventilation was stopped and the animal went through a sequence of normoxia, hypoxia and finally anoxia. During this sequence the heart failed. PCr decreased to undetectable levels within 10 min and Pi increased. The ATP concentration was unaffected until PCr was no longer observed, after which the ATP peak decreased, being no longer visible at 17 min. Similar results have been obtained by Neurohr et al. (1983a, b). These authors also looked at the natural abundance ^{13}C spectrum. Glycogen was only observed after the administration of 1-^{13}C glucose with the appearance of a resonance corresponding to the 1-^{13}C of glycogen. When the

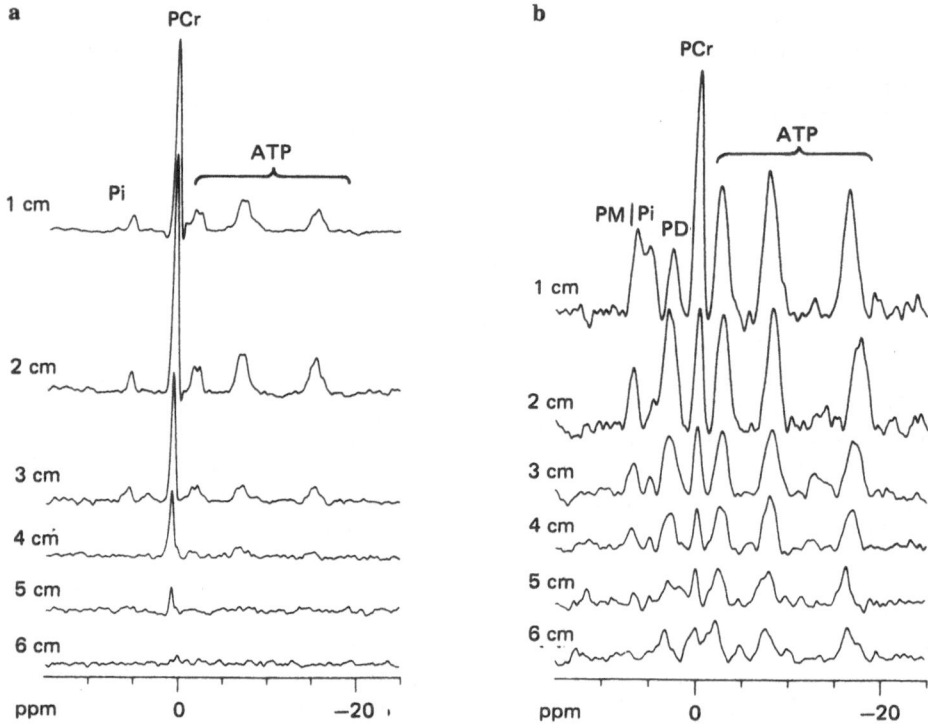

Fig. 4.7. **a, b** In vivo ³¹P SLIT DRESS. Spectra were recorded **a** with the lower leg draped over the surface coil showing the calf muscle and **b** with the head lying sideways on the coil above the left ear. Each complete spectral series was recorded in 5 and 10 min, respectively, with $T_r = 2$ s, and $n = 6$. Depths relative to the surface are indicated. *ATP*, phosphates of adenosine triphosphates; *PCr*, phosphocreatine; *PD*, phospodiester; *PM*, phosphomonoester; *Pi*, inorganic phosphate (Bottomley et al. 1985a).

heart was made ischaemic this resonance disappeared, to be replaced by one corresponding to the 3-¹³C lactate.

Nunnally and Bottomley (1981) were the first to apply a surface coil to in vivo heart studies. In this study they assessed the effectiveness of certain drugs in treating myocardial infarction. As described earlier, surface coils offer the advantage of spatial selectivity so that while ligation of the left anterior descending artery made little difference to the spectrum of a perfused rabbit heart obtained using a conventional coil arrangement (Hollis et al. 1978), suitable placement of a small surface coil (11 mm inner diameter) gave spectra showing large changes in PCr and adenine nucleotide concentrations after ligation. Nunnally and Bottomley treated a ligated heart with the drug verapamil used in the clinical management of cardiac arrhythmias and angina. After treatment there was a 220% increase in PCr levels, consistent with the hypothesis that verapamil produces coronary vasodilation and in so doing enhances collateral flow to the ischaemic region.

Studies using the purine inosine showed preservation of ATP concentration and myocardial function (Lewandowski and Devous 1985). Similar studies have been carried out in vitro on a number of cardioplegic solutions where spectroscopy provides a good method of assessing tissue viability (Bernard et al. 1985).

Recently, using the depth selective DRESS technique, ischaemia following coronary occlusion in a dog was demonstrated (Bottomley et al. 1985b).

It is clear that the diagnosis of ischaemic disease in both brain and heart is possible by ^{31}P NMR, requiring only spatial resolution of the area from where signals are to be obtained. The importance of such direct diagnosis, especially in myocardial disease, is obvious. Additionally, spectroscopy may prove an important technique in reviewing treatment regimens and assessing transplant viability.

Liver

The liver has proved a difficult organ for NMR investigation for several reasons and little ^{31}P work has been carried out in vivo.

Conventional vertical NMR magnets are not suitable for surface coil studies of liver due to the awkward position in which the animal is placed. Liver perfusion is more technically demanding than heart perfusion and investigation of liver was necessarily invasive. Recently, however, the advent of localisation methods has enabled studies to be carried out non-invasively (Canioni et al. 1983; Radda 1984).

Stevens et al. (1982), Sillerud and Shulman (1983) and Canioni et al. (1983) have used the surface coil technique to observe ^{13}C signals from liver. Stevens et al. (1982) and Sillerud and Shulman (1983) have detected glycogen by natural abundance and have reached the same conclusions from different approaches, primarily that glycogen is 100% visible by ^{13}C NMR — an interesting result considering the molecular size of glycogen. ^{13}C NMR may also be used to detect liver glycogen storage diseases by direct observation of glycogen (Stevens et al. 1983). It is, however, less likely to be practical in muscle, where glycogen levels are far lower.

Canioni et al. (1983), using the TMR technique, obtained liver ^{13}C spectra without recourse to surgery. They examined the adipose tissue of a number of rats receiving different diets: a normal balanced diet, a fat-free diet and a diet high in polyunsaturated fats. Differences in the respective spectra were observed, allowing identification of resonances corresponding to the olefinic carbons of poly-, and both poly- and mono-unsaturated fatty acid chains. The ratio of these peaks was used to determine the degree of polyunsaturation of the storage fats and the size of the total unsaturated pool.

Stevens et al. (1984) studied the metabolism of foreign compounds in the mouse liver using ^{19}F NMR and surface coils. The anticancer drug 5-fluorouracil was administered to tumour-bearing mice at twice or five times the recommended human dose. Catabolic products were seen in the liver and the time course of metabolism of the drug could be followed. In contrast, tumour tissue degraded the drug at a far slower rate and no breakdown products were observed; products of the anabolic pathway, the active metabolites, were seen (Fig. 4.5). The method opens up the way to the detection of a drug at its site of action rather than the reliance on plasma concentrations to infer correct dosage.

Recently Radda (1984) has examined the liver non-invasively in a human volunteer. Identification was carried out by the absence of phosphocreatine in the spectrum and the presence of a high phosphomonoester peak (fructose-1-phosphate) on fructose loading (Iles et al. 1981) of the volunteer. The authors describe the potential of spectroscopy to study disorders of carbohydrate metabolism, for example glucose-6-phosphatase deficiency.

Tumours

Localised spectroscopy provides the only method available to monitor tumour tissue. The promise of the technique is that it will open the way to the detection and diagnosis of malignancies, allow assessment of the metabolic and vascular state of tumours and predict their response to therapy. The area has been recently reviewed by Sostman et al. (1984).

Many workers have examined tumour implants in animals (for example Griffiths et al. 1981; Ng et al. 1982; Koetze et al. 1984). All describe similar spectra deriving from tumours of different histological origin. Interestingly, phosphocreatine is present in tumours of different histological type and has been observed in some fast growing hepatoma models. (J.R. Griffiths, personal communication). Changes with age and size are observed. Tumours appear to become less metabolically active as they grow. Large tumours often have little or no observable PCr and a reduced ATP; in contrast Pi and PME resonances are greatly elevated. Surprisingly pH is maintained near neutrality even when energetically the tumours show a gross hypoxic profile.

Changes in the tumour spectra are brought about by chemotherapy and irradiation (Ng et al. 1982).

After drug therapy (cyclophosphamide, BCNU or adriamycin) a diminution of signal/noise was observed with considerable change in the relative concentrations of high energy phosphates. This was magnified by a near reversal of the trend observed during growth, and most notably there was a decrease in the ATP/PCr ratio after treatment, with a noticeable elevation of PCr.

Following radiotherapy (14 Gy in mammary carcinoma) there was a transient rapid removal of PCr after 15 min before the general spectral pattern observed after drug therapy was seen, an increase in phosphocreatine associated with reduced tumour mass. The aim of such studies is to categorise the oxidation state of the tumours to identify reoxygenation — an important criterion in deciding on the timing of subsequent irradiation therapy to maximise response and minimise dose.

Hypothermia therapy has also been evaluated. Using RF heating, changes in energy metabolites were observed, with a lowering of the ATP/Pi ratios related to RF power (Naruse et al. 1985).

Conflicting data exist on the use of hyperglycaemia therapy. Griffiths et al. (1981) saw no change in the tumour spectrum on administration of glucose to rats carrying implants of murine tumours. Evelhoch et al. (1984), however, detected a decrease in intracellular pH of 0.45 units coinciding with a serum glucose of over 30 mM. In addition a 50% fall in PCr was observed. For a number of reasons, it may be unreasonable to expect reproducible results in tumour studies. Different tumour lines are likely to vary in their ability to cope with an increased glucose load. The state of vascularisation will undoubtedly affect both glucose delivery, its fate, and possible lactate export. The presence of hypoxic cell populations may also give rise to misleading results.

Changes have also been observed in hormone responsive tumours.

In a study of mammary tumour (Rodrigues et al. 1985), removal of oestrogen, the stimulus for growth, by ovariectomy led to a reversal in the ^{31}P pattern observed during growth and a diminution in tumour size. Studies to evaluate anti-oestrogens continue and this may prove an important area in evaluating response to therapy.

The effects of the polypeptide hormones VIP and TRH on the metabolism of a pituitary tumour line were studied by Prysor Jones et al. (1984). TRH is able to cause release of the hormone prolactin while VIP is unable to do so. However, treatment with TRH subsequent to VIP leads to a larger release of prolactin than can be elicited with TRH alone. Breakdown of PCr and rise in Pi is seen by NMR after either of the treatment regimes but is more marked with VIP; pH changes are also seen in this case. While no interpretation could be made, the observations were interesting in that they confirmed an underlying mechanism to VIP priming and provided the first example of NMR investigation of the secretion process.

Studies of tumour metabolism using nuclei other than ^{31}P have been restricted to ^{19}F studies of 5-fluorouracil uptake into tumour implants in mice (Stevens et al. 1984) and have been discussed in the previous section.

To date, few human tumours have been studied in situ by NMR spectroscopy.

Griffiths et al. (1983) investigated a rhabdomyosarcoma on the dorsum of the hand. Unique spectra were obtained which contrasted with the adjacent muscle. However, after chemotherapy, they were unable to see spectral changes.

Ross et al. (1984) have examined tumour-bearing kidneys after removal. It proved possible by perfusion to reinstate the normal O_2 consumption, glomerular filtration and NMR spectra of the kidney. Observation of the tumour showed that during hypoxia the tumour was able to conserve its ATP better than the normal kidney. In a number of tumours they identified additional resonances at 4 ppm and assigned this to a pool of phosphate at low pH. Interestingly, this resonance disappeared during hypoxic shock.

Chance et al. (1985) studied two infants with enlarged livers due to the presence of neuroblastoma. They identified the tumour on the basis of a characteristic ^{31}P NMR spectrum, notably an elevated phosphomonoester resonance. In following the tumour over time, changes in the spectra were observed which could be correlated with clinical observations. Unfortunately, the study suffered from the inability to spectroscopically localise the tumour.

Unfortunately, of the tumour studies to date, there is considerable similarity between tumour and other tissues and it remains unlikely that, for example, glioma could be differentiated from brain tissue on the basis of ^{31}P spectral data. In combination with proton imaging, however, spectroscopy promises to be a useful tool in cancer therapy.

Conclusion

Spectroscopy has advanced over the last decade: it provides a valuable technique for understanding the underlying metabolic causes of disease as well as permitting direct identification. As we have seen, new pulse techniques have allowed easy observation of ^{13}C and ^{1}H nuclei. The analytical accuracy of the technique has greatly advanced but the difficulty in obtaining absolute quantitation remains a weakness.

There are two directions in which spectroscopy may develop: either to identify disease in a manner akin to clinical biochemistry and to monitor therapy, or to use its unique properties to understand more fully the physical basis of disease processes, for example the considerable work already carried out in studying muscle physiology. Given the cost of a simple enzyme assay it would be naïve to assume that spectroscopy will replace a CK (creatine kinase) estimation in identifying ischaemic damage; however, as a monitor of therapy it may have no equal. This is best seen in the field of oncology where currently the ability to stage and follow therapy non-invasively is limited.

We remain on the threshold: with spectroscopy, the next year will show whether we will cross into the world of remote monitoring of disease.

References

Ackerman JJH, Grove TH, Wong GG, Gadian DG, Radda GK (1980) Mapping of metabolites in whole animals by ^{31}P NMR using surface coils. Nature 283: 167–170

Ackerman JJH, Berkowitz BA, Devel RK (1984) Phosphorus [31]NMR of rat brain in vivo with bloodless perfluorocarbon in the perfused rat. Biochem Biophys Res Commun 119: 913–919

Arnold DL, Bore PJ, Radda GK, Styles P, Taylor DJ (1984) Excessive intracellular acidosis of skeletal muscle on exercise with a patient with a post viral exhaustion/fatigue syndrome. Lancet I: 1367–1369

Aue WP, Muller S, Cross TA, Seelig J (1984) Volume-selective excitation. A novel approach to topical NMR. J Magn Reson 56: 350–354

Balaban RS, Kantor HL, Ferretti JA (1983) In vivo flux between PCr and adenosine triphosphate determined by 2 dimensional ^{31}P NMR. J Biol Chem 258: 12787–12789

Behar KL, Den Hollander JA, Stromskime et al. (1983) High resolution H NMR study of cerebral hypoxia in vivo. Proc Natl Acad Sci USA 81: 2517–2519

Behar KL, Rothman DL, Shulman RG, Pettroff OAC, Pritchard JW (1984) Detection of cerebral lactate in vivo during hypoxaemia by ^{1}H NMR at low field strength (1.9 T). Proc Natl Acad Sci 81: 2517–2519

Bendall MR, Gordon RE (1983) Depth and refocusing pulses designed for multi-purpose NMR with surface coils. J Magn Reson 53: 365–385

Bernard M, Menasche P, Canioni P et al. (1985) ^{31}P NMR study of the perfused heart; protective effects of cardio-plegic solutions. Proceedings of the 4th Meeting of the Society of Magnetic Resonance in Medicine. SMRM, Berkeley, CA, pp 434–444

Bottomley PA, Hart HR, Edelstein WA et al. (1983) NMR imaging/spectroscopy system to study both anatomy and metabolism. Lancet II: 273–274

Bottomley PA, Smith LS, Leve WM, Charles C (1985a) Slice interleaved depth resolved surface coil (SLIT DRESS) for rapid ^{31}P NMR in vivo. J Magn Reson 64: 347–351

Bottomley PA, Herfkens R, Smith LS et al. (1985b) Non-invasive detection and monitoring of regional myocardial ischaemia in situ using depth resolved ^{31}P NMR spectroscopy. Proc Natl Acad Sci USA 82: 8747–8751

Cady EB, De L, Costello AM et al. (1983) Non-invasive investigation of cerebral metabolism in newborn infants by ^{31}P NMR. Lancet I: 1059–1062

Canioni P, Alger JR, Shulman RG (1983) Natural abundance ^{13}C NMR spectroscopy of liver and tissue of living rat. Biochemistry 22: 4974–4980

Chance B, Elleff S, Leigh JS (1980) Non-invasive non-destructive approach to cell bioenergetics. Proc Natl Acad Sci USA 77: 7430–7434

Chance B, Elleff S, Leigh JS, Sokolow D, Sapega A (1981) Mitochondrial regulation of PCr/Pi ratio in exercising human limbs. Gated ^{31}P NMR study. Proc Natl Acad Sci USA 78: 6714–6718

Civan MM, Lin L-E, Peterson-Yantorno K, Taylor J, Deutsch C (1984) Intracellular pH of perfused single frog skin; combined ^{19}F and ^{31}P analysis. Am J Physiol 247: (Cell Physiol 16) C506–C510

Cresshull ID, Dawson MJ, Edwards RHT et al. (1981) Human muscle analysed by ^{31}P NMR in intact subjects. J Physiol 317: 18

Edwards RHT, Dawson JM, Wilkie DR, Gordon RE, Shaw D (1982) Clinical use of NMR in the investigation of myopathy. Lancet I: 725–732

Evanochko WT, Ng TC, Lilly MB et al. (1983) ^{31}P NMR study of murine mammary 16/C adenocarcinoma and its response to chemotherapy, x-radiation and hyperthermia. Proc Natl Acad Sci USA 80: 334–338

Evelhoch JL, Sapareto SA, Jick DEL, Ackerman JJH (1984) In vivo metabolic effects of hyperglycaemia in murine RIF tumour; a ^{31}P NMR investigation. Magn Reson Med 1: 209–216

Gadian DG, Radda GK, Ross BD et al. (1981) Examination of a myopathy by ^{31}P NMR. Lancet II: 774–775

Gadian DG, Williams SR, Proctor E, Cady EB, Gardiner M (1986) Neurometabolic effects of an inborn error of amino acid metabolism demonstrated in vivo by ^{1}H NMR. Magn Reson Med 3: 150–156

Gordon RE, Hanley PE, Shaw D et al. (1980) Localisation of metabolites in animals using ^{31}P topical NMR. Nature 287: 367–370

Griffiths JR, Stevens AN, Iles RA, Gordon RE, Shaw D (1981) ^{31}P NMR investigation of solid tumours in the living rat. Biosci Rep 1: 319–325

Griffiths JR, Cady EB, Edwards RHT, McCready VR, Wilkie DR, Wiltshaw E (1983) ^{31}P NMR studies of a human tumour in situ. Lancet I: 1435–1436

Grist TM, Hyde JS (1985) Resonators for in vivo ^{31}P NMR at 1.5 T. J Magn Reson 61: 571–578

Grove TH, Ackermann JJH, Radda GK, Bore PJ (1980) Analysis of rat heart in vivo by phosphorous NMR. Proc Natl Acad Sci USA 77: 299–302

Gupta RK, Gupta P (1982) Direct observation of resolved resonances from intra- and extra-cellular ^{23}Na ions in NMR studies. J Magn Reson 47: 344–350

Hilal SK, Maudsley AA, Simon HE et al. (1983) In vivo NMR imaging of tissue sodium in the intact cat before and after acute cerebral stroke. AJNR 4: 245–249

Hollis DP, Nunnally RL, Taylor GT, Weisfeldt ML, Jacobus WE (1978) Phosphorus nuclear magnetic resonance studies of heart physiology. J Magn Reson 29: 319–330

Hore PJ (1983) Solvent suppression in Fourier transform nuclear magnetic resonance. J Magn Reson 55: 283–300

Iles RA, Stevens AN, Griffiths JR (1982) NMR studies of metabolites in living tissues. Prog NMR Spectrosc 15: 49–200

Iles RA, Griffiths JR, Stevens AN, Gadian DG, Porteous R (1981) Effects of fructose on the energy metabolism and acid base status of the perfused starved rat liver. Biochem J 192: 191–202

Koeze TH, Lantos PL, Iles RA, Gordon RE (1984) In vivo nuclear magnetic resonance spectroscopy of a transplanted brain tumour. Br J Cancer 49: 357–361

Lewandowski ED, Devons MD (1985) Inosine preserves ATP and function is ischaemic and reperfused myocardium. Proceedings of the 4th Meeting of the Society of Magnetic Resonance in Medicine. SMRM, Berkeley, CA, pp 504–505

Maris JM, Evans AE, McLaughlin AC et al. (1985) ^{31}P NMR spectroscopy investigating of human neuroblastoma in situ. N Engl J Med 312: 1500–1505

Maudsley AA, Hilal SK, Simon HE, Wittenkoeks S (1984) In vivo spectroscopic imaging with [31]P. Radiology 153: 745–750

Naruse S, Horikawa Y, Tanaka C, Hirakawa K, Nishikawa H, Watari H (1984) In vivo measurement of energy metabolism and the concomitant monitoring of electroencephalogram in experimental cerebral ischaemia. Brain Res 296: 370–372

Naruse S, Horikawa Y, Tanaka C, Higushi T, Hirakawa K, Nishakawa H (1985) RF hyperthermia and simultaneous monitoring of its effect on tumour using surface coil ^{31}P NMR spectroscopy. Proceedings of the 4th Meeting of the Society of Magnetic Resonance in Medicine, SMRM, Berkeley CA, pp 512–513

Neurohr KJ, Barret EJ, Shulman RG (1983) In vivo carbon — 13 NMR studies of heart metabolism. Proc Natl Acad Sci USA 80: 1603–1607

Neurohr KJ, Gollin G, Barret EJ, Shulman RJ (1983) In vivo ^{31}P NMR studies of myocardial high energy phosphate metabolism during anoxia and recovery. FEBS Lett 159: 207–210

Ng TC, Evanochko WT, Hiramoto RN et al. (1982) ^{31}P NMR spectroscopy of in vivo tumours. J Magn Reson 49: 271–286

Nunnally RL, Bottomley PA (1981) Assessment of pharmacological treatment of myocardial infarction. Science 211: 177–180

Oberhaensli R, Galloway G, Taylor D, Bore P, Radda G (1985) Functional assessment of human liver metabolism by ^{31}P MRS. Proceedings of the 4th Meeting of the Society of Magnetic Resonance in Medicine, SMRM, Berkeley, CA, pp 520–521

Ordige RJ, Connelly A, Loman JAB (1986) Image selected in vivo spectroscopy (ISIS). A new technique for spatially selective NMR spectroscopy. J Magn Reson 66: 283–294

Pettegrew JW, Minshew NJ, Diehl J, Smith T, Kopp SJ, Glonek T (1983) Anatomical considerations for interpreting typical ^{31}P NMR, Lancet II: 913

Pritchard JW, Alger JR, Behar KL, Petroff OAC, Schulman RG (1983) Cerebral metabolic studies in vivo by ^{31}P NMR. Proc Natl Acad Sci USA 80: 2748–2751

Prysor-Jones RA, Silverlight JJ, Jenkins JS, Stevens AN, Rodrigues JL, Griffiths JR (1984) VIP enhances TRH-stimulated prolactin secretion of pituitary tumours; studies with ^{31}P NMR. FEBS Lett 177: 71–75

Radda GK (1984) Clinical studies by ^{31}P NMR spectroscopy. Proceedings of the 4th Meeting of the Society of Magnetic Resonance in Medicine, SMRM, Berkeley, CA, pp 605–608

Radda GK, Bore PJ, Gadian DG, Styles P, Taylor D, Morgan-Hughes J (1982) ^{31}P NMR examination of two patients with NADH CoQ reductase deficiency. Nature 295: 608–609

Radda GK, Bore PJ, Ragagpoulan B (1983) Clinical applications of ^{31}P NMR spectroscopy. Br Med Bull 40: 155–159

Rodrigues JL, Stevens AN, Wilkinson J, Coombes C, Griffiths JR (1985) Chemically induced mammary tumours. A ^{31}P NMR study of growth therapy. Biochem Soc Trans 13: 887

Ross BD, Marshall V, Smith M, Bartlett S, Freeman D (1984) Monitoring response to chemotherapy of intact human tumours by ^{31}P NMR. Lancet I: 641–646

Ross BD, Radda GK, Gadian DG, Rocker G, Esiri M, Falconer-Smith J (1981) Examination of a case of suspected McArdle's syndrome by ^{31}P NMR. N Engl J Med 304: 1338–1342

Rothman DL, Behar KL, Hetherington HP, Den Hollander JA, Bendall MR, Shulman RJ (1985) ^{1}H observe ^{13}C decouple spectroscopic measurements of lactate and glutamate in the rat brain in vivo (abstr). Proc Natl Acad Sci USA 82: 1633–1637

Shoubridge EA, Briggs RW, Radda GK (1982) Saturation transfer measurements of the steady state rates of creatine kinase and ATP. FEBS Lett 140: 288–292

Sillerud LO, Shulman RG (1983) Structure and metabolism of mammalian liver glycogen monitored by ^{13}C NMR. Biochemistry 22: 1087–1094

Sostman HD, Armitage IM, Fischer JJ (1984) NMR in lancer, high resolution spectroscopy of tumours. Magn Reson Imag 2: 265–278

Smith GA, Hesketh TR, Metcalfe JC, Feeny J, Morris PG (1983) Intracellular calcium measurements by ^{19}F NMR of fluorine labeled chelators. Proc Natl Acad Sci USA 80: 7178–7184

Stevens AN, Iles RA, Morris PG, Griffiths JR (1982) Detection of glycogen in a glycogen storage disease by ^{13}C NMR. FEBS Lett 150: 489–493

Stevens AN, Morris PG, Iles RA, Sheldon PW, Griffiths JR (1984) 5 fluorouracil metabolism monitored in vivo by ^{19}F NMR. Br J Cancer 50: 113–117

Styles P, Scott CA, Radda GK (1985) A method for localising high resolution NMR spectra from human subjects. Magn Reson Med 2: 402–409

Taylor DJ, Bore PJ, Styles P, Gadian DG, Radda GK (1983) Bioenergetics of intact human muscle: a ^{31}P NMR study. Mol Biol Med 1: 77–94

Tofts PS, Cady EB, Delpy DT et al. (1984) Surface coil NMR spectroscopy of the brain. Lancet I: 459

Williams SR, Gadian DG, Proctor E et al. (1985) ^{1}H NMR studies of muscle metabolites in vivo. J Magn Reson 63: 406–412

Wyrwicz AM, Pszenny MH, Schofield JC (1984) Observations of fluorinated anaesthetics in rabbit brain by ^{19}F NMR. Magn Reson Med 1: 275–276

5. Functional Magnetic Resonance Imaging of the Heart and Large Vessels

R.E. Coupland and P. Mansfield

Introduction

Imaging times using planar imaging techniques including projection reconstruction and two-dimensional Fourier transformation methods are comparatively long, typically 2–5 min: in consequence patient throughput is reduced. Problems also arise with regard to involuntary movements, e.g. of viscera, which degrade images. Gating procedures, especially in the heart, can to some extent overcome the problems of movements that would otherwise blur or degrade the image.

Since 1976–77 a new approach to the problem of imaging of moving organs has been developed. This allows the encoding and observing of all regions in a cross-sectional slice in a single experiment that lasts only a few tens of milliseconds. The technique is commonly referred to as echo planar imaging (EPI) (Mansfield 1977; Mansfield and Morris 1982).

Method

The EPI method has been fully described elsewhere (Mansfield 1977; Mansfield and Morris 1982) so only a brief description is given here.

Initially a tailored RF pulse and magnetic field gradient G_z are used in a slice selection procedure (Garroway et al. 1974) to excite nuclear spins within a predetermined slice. The signal response is then recorded. During this recording process two further gradients G_x and G_y are applied which spatially encode the spins within the selected slice along the x- and y-axis. These gradients have different time dependencies resulting in a unique assignment of frequencies to each pixel in the two-dimensional image array. Upon Fourier transformation of the signal response, these frequencies are unambiguously unravelled and used to identify each pixel amplitude: these are then properly ordered into an image format for display.

Since this process is performed in a one shot experiment the imaging time is extremely fast. Typical imaging times currently used are 32 ms and 64 ms.

If the imaging process is repeated a delay is necessary between successive shots to allow the spins to repolarise. Repolarisation occurs with a time constant T_1, the spin lattice relaxation time. Because of T_1 effects, EP images are T_1 weighted, rather than pure spin density maps. This can be an advantage, since T_1 weighting tends to increase image contrast.

Imaging Details and Procedure

The apparatus used to produce the images presented here comprised a resistive magnet with a field strength of 0.094 T, corresponding to an RF frequency of 4.0 MHz for protons, and the EPI probe access was 22 cm with an object field of 19.2 mm. The image field was represented by a 64×64 matrix with pixel resolution of 3 mm. The selective slice thickness was 7 mm. The images were linearly interpolated to 256 arrays for photographic recording. Magnetic field gradients used were $G_x = 0.13$ mTm^{-1}, $G_y = 8.0$ mTm^{-1} and $G_z = 3.48$ mTm^{-1}. The maximum switching rates for the x, y and z gradients were 16 Tm^{-1}s^{-1}, 512 Tm^{-1}s^{-1} and 109 Tm^{-1}s^{-1} respectively. The estimated maximum duration over which these rates applied was 20 μs. The selective radio frequency (RF) pulse power has a peak of 8 W delivered to the probe with an average power of 2 W over a 3.5-ms period.

For EPI studies of the heart and great vessels of infants and young children, 32 contiguous slices were imaged starting at the level of the shoulders and ending in the mid-abdomen. At each level 16 images were taken during sequential cardiac cycles. The repetition period was adjusted to be slightly greater than the length of the cardiac cycle as measured from an ECG monitor, i.e. $\tau_E = \tau_C(1 + 1/m)$ where m is the number of frames required to uniformly span the cardiac cycle, and τ_C = heart period (1/rate). In most cases the time taken to obtain the 16 frames was 8 s. During some scans the ECG trace was simultaneously recorded in order to correlate particular images with the various phases of the cardiac cycle. In others, the images were triggered following a variable stepped delay with respect to the R wave.

This imaging procedure resulted in the production of 512 images in about 4.5 min. These constituted a four-dimensional data set. From this movie sequences or still images were reconstructed in order to obtain cross-sectional, coronal or sagittal views of the patient.

The data sampling for each snap shot image was 64 ms. Further data manipulation and display screen update take an additional 200 ms, thus restricting the repetition rate to about four frames per second in the immediate viewing or real time imaging mode. Using a visual playback mode instead of immediate viewing the repetition rate can be increased to ten frames per second. Movie sequences constructed from arbitrary planes through the subject may be viewed subsequently.

Results and Conclusions

Figure 5.1 shows several contiguous transectional views through the thorax of an 8-month-old child. (Chrispin et al. 1986a, b). These demonstrate the variations in position of the chambers in a normal heart as one proceeds from mid-heart, Fig. 5.1a, through the thorax to the base of the heart, Fig. 5.1f. All images were taken at mid-systole. Owing to the spiral course of the outflow tracts from right and left ventricles these change in position relative to each other as one passes sequentially through aorta and pulmonary trunk to aortic vestibule and infundibulum of right ventricle and finally to the ventricular cavities. Atria and pulmonary vessels are visible in association with the posterior aspects of the heart. The superior vena cava can be identified to the right.

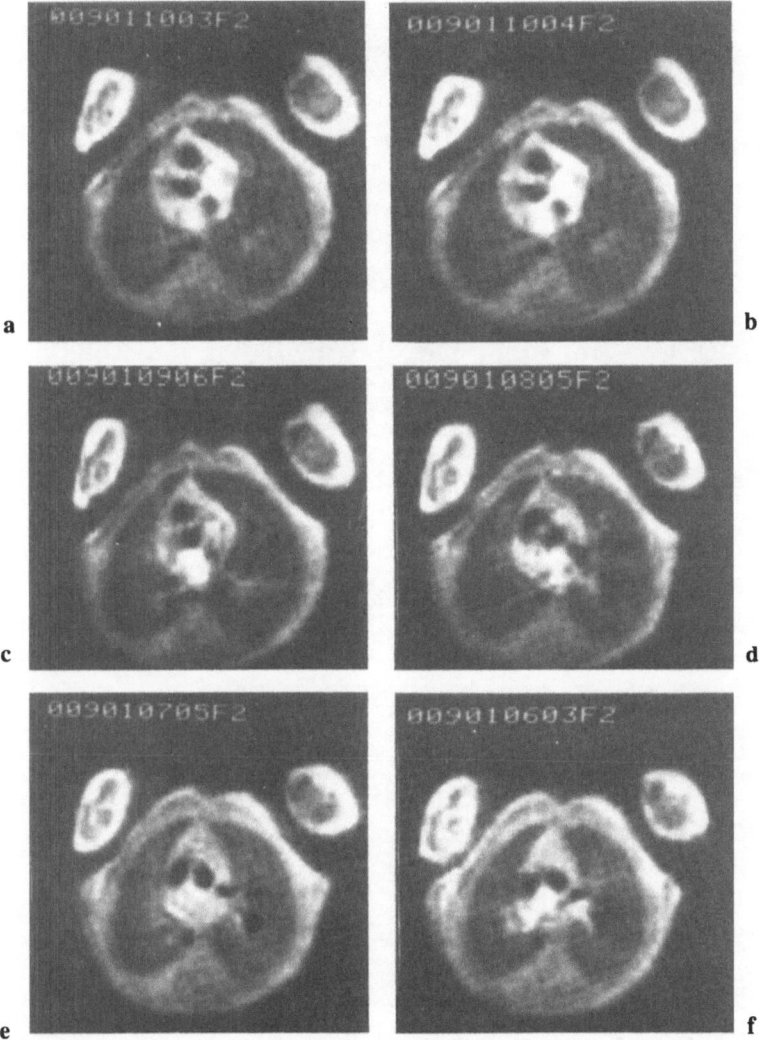

Fig. 5.1a–f. Contiguous transectional snap-shot images through the thorax of an 8-month-old child with a normal heart. **a** corresponds to mid-heart proceeding cranially to the base of heart, **f**. All images are taken at mid-systole, where chambers and vessels show as dark zones within the heart mass. The left side of the patient is *left* and the anterior aspect at the *bottom* in all pictures (Chrispin et al. 1986a).

The blood within the chambers of the heart gives a low intensity image while in rapid motion and a bright image when at or near standstill. Although the information available by virtue of this variation has still not been fully analysed, clearly it is in principle possible to make use of this for the determination of blood flow.

From the same data set, animated sagittal sections can be produced which show the ejection of blood from the left ventricle into the aorta and its passage through the

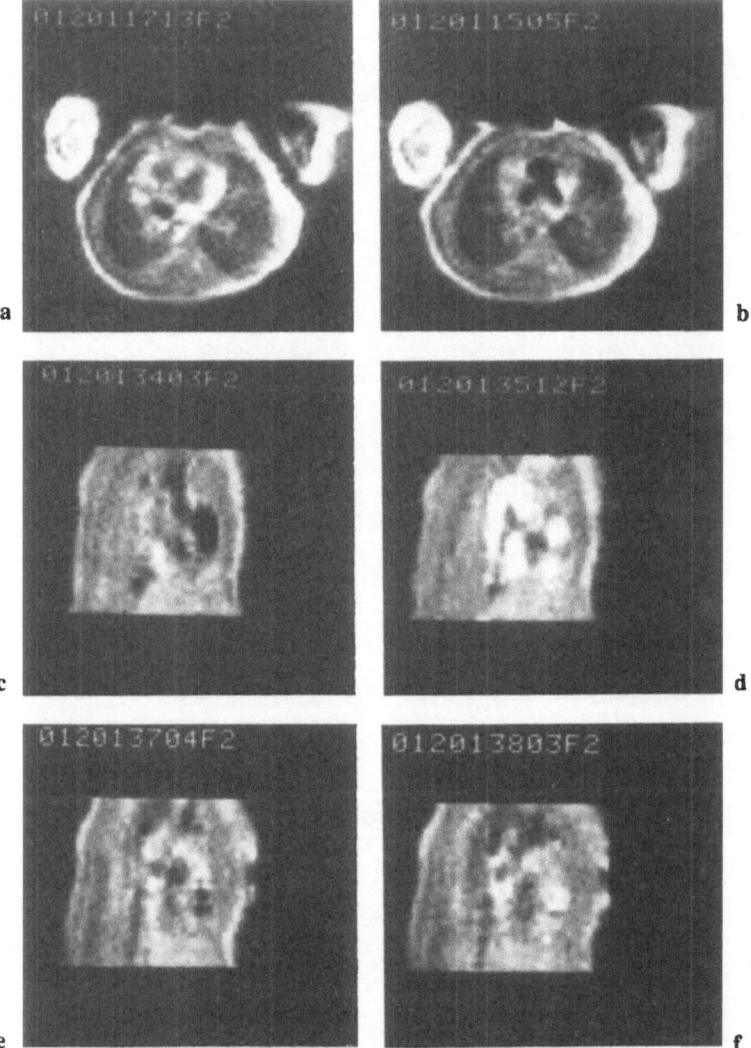

Fig. 5.2a–f. Echo-planar images from a patient with simple transposition of the major vessels of the heart. **a** is a transection showing chambers of the heart; **b** is 12 mm cephalad and shows the bifurcation of the pulmonary artery; **c** and **d** are adjacent sagittal views showing the connections of right ventricle to ascending aorta (**c** is at systole and **d** at diastole); **e** and **f** are views further left taken at systole. In all sagittal views the posterior aspect is *left* (Chrispin et al. 1986a).

descending thoracic and abdominal aorta. The origin of the coeliac axis artery is clearly visible, as are two of its branches. All these structures show variations in signal intensity according to the rate of blood flow during the various phases of the cardiac cycle.

Figure 5.2 shows some images through the thorax of a 3-month-old infant born with transposition of the great vessels. Figures 5.2a and 5.2b are examples of transaxial sections from a set of 32 contiguous slices. An enlarged atrium with a large atrial septal defect is visible in more cranial cuts while the ventricles and muscular part of the interventricular septum are clearly visible in more caudal sections. Figures 5.2c-f are examples of sagittal sections constructed from the four-dimensional data set and clearly show the connection between right ventricle and ascending aorta. Animated sagittal and coronal plane presentation of the data shows blood being ejected from the right ventricle into the aorta and from the left ventricle into pulmonary trunk and arteries.

Other cases studied include a child with right heart hypoplasia.

The congenital defects present in these cases had been established clinically by using cardiac catheterisation and angiography, echo-cardiography, ECG and chest radiography as appropriate. However, the clinical diagnosis was not made known to those evaluating the EP images until after a report had been made.

Echo planar imaging provides real time movie sequences of moving organs in situ in a totally non-invasive way. Within limits set by the probe size, EPI allows movements of the patient to be immediately screened during the imaging procedure, thus allowing optimal image plane selection.

References

Chrispin A, Small P, Rutter N et al. (1986a) Transectional echo planar imaging of the heart in cyanotic congenital heart disease. Paediatr Radiol 16

Chrispin A, Small P, Rutter N et al. (1986b) Echo planar imaging of normal and abnormal connections of the heart and great arteries. Paediatr Radiol 16

Garroway AN, Grannell PK, Mansfield P (1974) Image formation in NMR by a selective irradiative process. J Phys C Solid State Phys C7:L437–482

Mansfield P, Morris PG (1982) NMR imaging in biomedicine. Academic Press, New York

Mansfield P (1977) Multiplanar image formation using NMR spin echoes. J Phys C Solid State Phys C10:L55–58

6. Functional Studies of the Cardiovascular System Using Magnetic Resonance Imaging

S. Richard Underwood

Introduction

Magnetic resonance allows high resolution tomographic imaging of most organs in the body. Static organs such as the brain and spinal cord provide excellent images and these are proving to be clinically valuable. Images of the heart may be produced at any point of the cardiac cycle using electrocardiographic gating, and flowing blood usually gives no signal. This provides high contrast between blood and myocardium without the use of contrast agents and allows the demonstration of cardiac anatomy (Steiner et al. 1983; Higgins et al. 1985a, b), and both global (Longmore et al. 1985; Byrd et al. 1985; Friedman et al. 1985) and regional ventricular function (Fisher et al. 1985). Although other non-invasive techniques are able to provide similar information, magnetic resonance is particularly versatile, requiring only a single study to show both high resolution anatomy in any plane, and accurate function. Its wider potential, however, lies in the ability to provide functional information other than that of moving anatomy. Biochemical effects can be studied through the relaxation parameters T_1 and T_2, and through magnetic resonance spectroscopy which is discussed in detail in Chap. 4. Magnetic resonance angiography forms images of the vascular system and flow measurements allow accurate mapping of the blood flow within it. Many aspects of cardiovascular function may therefore be studied, and magnetic resonance will find an important place in the future management of cardiovascular disease.

Imaging Techniques

Because the heart is a moving organ, techniques used for other organs are not always appropriate. Some of the modifications are discussed below.

Cardiac Gating

Electrocardiographic gating is essential for cardiac imaging because without it the heart will be at a different part of the cardiac cycle for each acquisition pulse sequence. It is not a simple matter to obtain a recognisable electrocardiogram during imaging. A fibre optic cable is usually used to transmit the signal from the magnetic field, since conducting wires may act as radio aerials and introduce noise. Another problem is the electrical potential induced by the motion of the heart within the magnetic field which leads to extra waves on the electrocardiogram and in high fields may make recognition of the R wave difficult. The changing magnetic field gradients of the acquisition pulse sequences induce very large potentials and it is necessary to clamp the ECG voltage during the sequences to avoid triggering from these.

In patients with an irregular rhythm such as atrial fibrillation, images can still be acquired although the diastolic images in particular are of reduced quality because of the varying filling periods of each cycle.

Respiratory gating improves image quality but acquisition time is unacceptably prolonged. Techniques exist to reduce respiratory artefact without prolonging acquisition and these are valuable in the thorax as well as the abdomen (Bailes et al. 1985).

Pulse Sequences

In magnetic resonance imaging it is usual to vary acquisition parameters to alter contrast between tissues, and to improve differentiation between them, but for the heart the choice is more limited, since the repetition time of the sequence must be the same as the RR interval or a multiple of it. The relatively long inversion recovery sequences cannot acquire an image during systole, but late diastolic images are possible. Because of this, the shorter spin echo sequence is more usual, although with short echo times (less than 30 ms) a signal may be obtained from relatively static blood within the cardiac chambers and natural contrast between the blood and myocardium may be lost.

Imaging Planes

One great strength of magnetic resonance is its ability to image in any plane and this is particularly important for the heart which lies at an oblique angle within the chest. Anatomy may be easily demonstrated in coronal, sagittal and transverse planes, but for the assessment of left ventricular function, planes parallel and perpendicular to the long axis of the left ventricle are preferable (Akins et al. 1985; Underwood et al. 1985a; Feiglin et al. 1985a).

Imaging Protocols

End-diastolic and end-systolic images are necessary for the measurement of ventricular function, and each must be acquired in a number of sections covering the whole heart. Six 10-mm sections are usually sufficient for the left ventricle although up to 12 may be necessary for the right ventricle, but diastole may be acquired in one level with systole in another and imaging time is therefore halved. A complete study may take between 30 and 40 min depending upon the heart rate and the number of levels required.

An alternative approach is to use a multislice sequence to acquire a number of different sections, each one at a different part of the cardiac cycle. The acquisition is repeated four times in order to acquire each section at each part of the cardiac cycle (Fisher et al. 1985).

Images of the same section at multiple parts of the cardiac cycle may be displayed as a cine loop. The human eye is particularly good at integrating the information presented in this way, and all cardiologists are accustomed to assessing left ventricular function from cine angiograms. In the field of nuclear cardiology there is already a large experience of obtaining functional information from such cine loops by the creation of functional images, and most of these are applicable to magnetic resonance. The images may be acquired separately, but acquisition time is reduced if they are acquired with a cine sequence (Jenkins et al. 1985; Feiglin et al. 1985b). Echo planar imaging allows very rapid acquisition and the images are almost displayed in real time (Ordidge et al. 1982). Machines that can do this are not yet available commercially; when they are, real time magnetic resonance imaging and the ability to image during stress will add greatly to the value of the technique.

Clinical Applications

Global Ventricular Function

A first step in assessing global ventricular function is the measurement of volume at end-diastole and end-systole, from which stroke volumes and ejection fractions may be calculated. Volumes can be measured accurately by summing the areas of the chamber in multiple contiguous sections, and it has been demonstrated that these volumes are accurate to within approximately 2% (Longmore et al. 1985). Table 6.1 shows mean values in 20 normal subjects with standard deviations (SD) for left (LV) and right ventricular (RV) end-diastolic volume (EDV), stroke volume (SV) and ejection fraction (EF). The true value of the stroke volume ratio (LVSV/RVSV) is 1, since left and right ventricular outputs have to be the same over the period of imaging (Longmore et al. 1985).

The multislice method of measuring ventricular volume is accurate but time consuming. A simpler approximation for the left ventricle may be obtained from an oblique slice containing the long axis of the ventricle (Fig. 6.1). The area (A) and length (L) of the cavity are measured and, assuming the ventricle to be an ellipsoid of revolution, its volume is given by the formula: $V = 8 \times A^2 / 3\pi \times L$

Table 6.1. Measured values for normal ventricular volumes

	LVEDV	LVSV	LVEF	RVEDV	RVSV	RVEF	LVSV/RVSV
Mean	132 ml	81 ml	61%	152 ml	80 ml	53%	1.01
SD	14.7	16.2	7.2	20.9	16.4	7.4	0.067

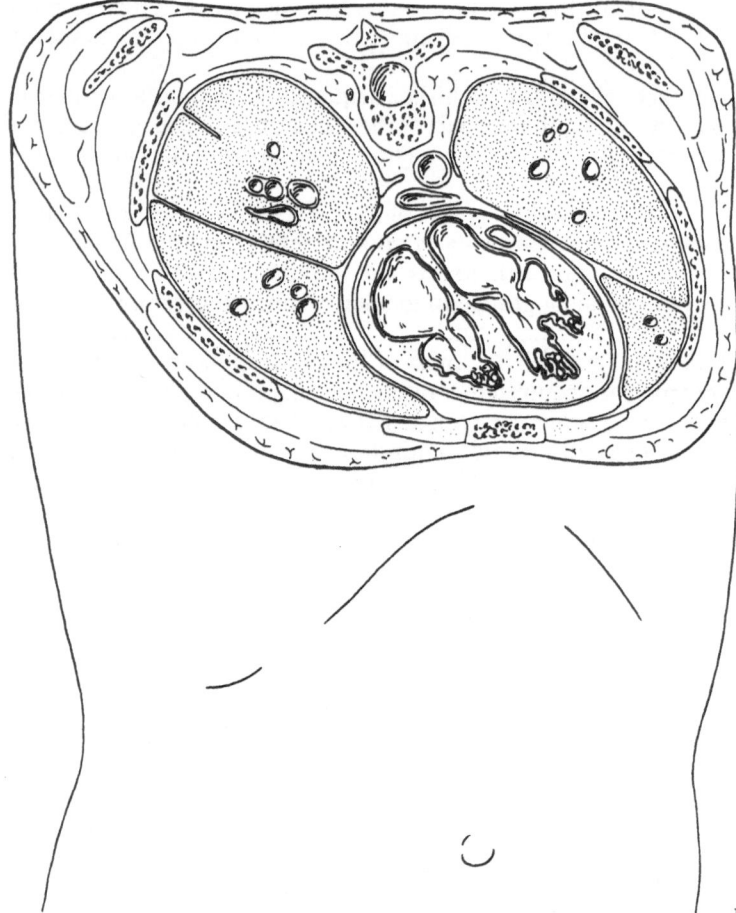

Fig. 6.1a.

Regional Ventricular Function

Measurements of ventricular wall motion and thickness are parameters of regional function. Wall motion can be assessed by a comparison of diastolic and systolic ventricular contours and wall thickness is measured directly from the images, although caution is required if the imaging plane is not perpendicular to the wall. Both parameters are sensitive but not specific indicators of previous infarction, since non-coronary myocardial disease also leads to abnormalities (Underwood et al. 1986a). Figure 6.2 shows a patient with an apical left ventricular aneurysm. The left ventricle is dilated with an ejection fraction of 23% and the apical myocardium is thin and dyskinetic.

A full description of regional function cannot be obtained from single tomographic sections but multiple sections present the problem of interpretation of a large number of images. Techniques already exist to reduce a number of tomographic images to a single image displaying three-dimensional data, and it is possible to display all parts of the ventricle from a series of short axis sections (Underwood et al. 1985b).

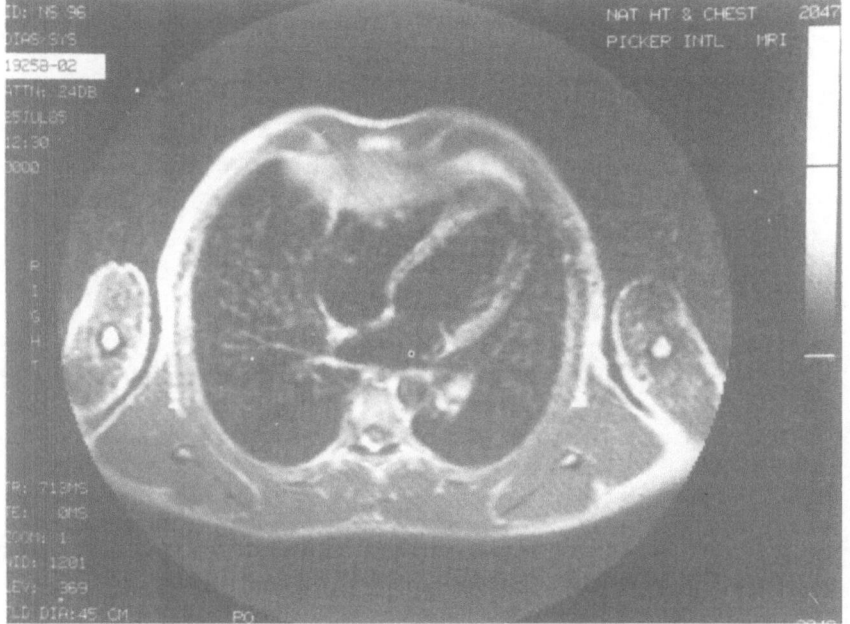

b

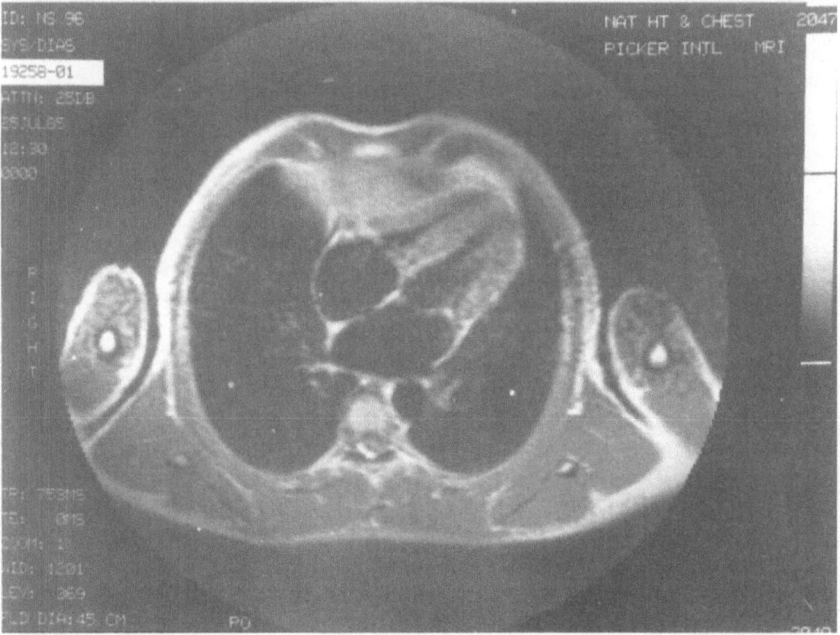

c

Fig. 6.1. a Line drawing to illustrate a transverse section tilted towards the coronal plane to give a four chamber section of the heart containing the long axis of the left ventricle. **b, c** Images in this plane at end-diastole (**b**) and end-systole (**c**). Note the symmetrical thickening and contraction of the left ventricular myocardium in this normal subject. The left ventricular volume can be calculated from the area and length of the cavity.

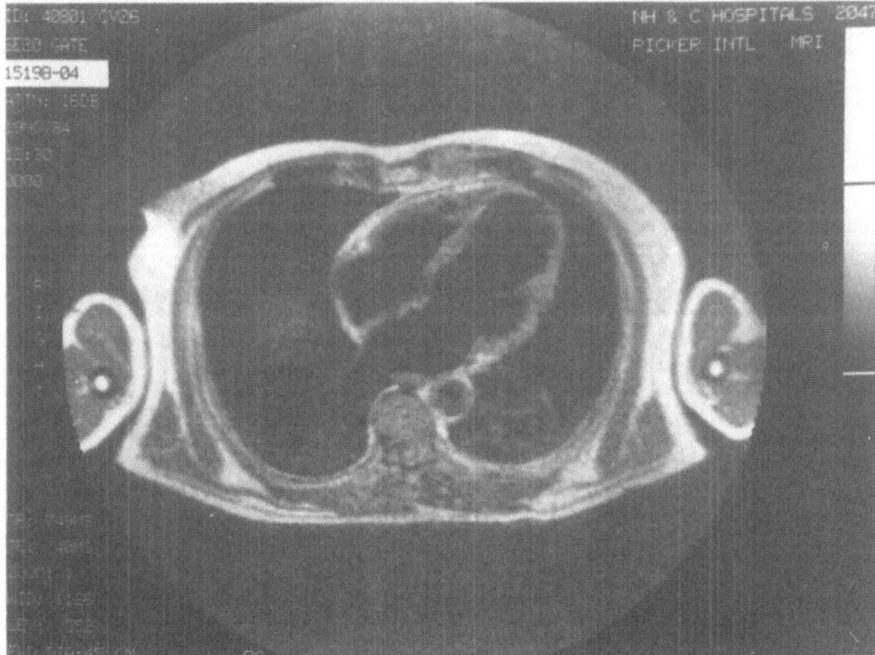

a

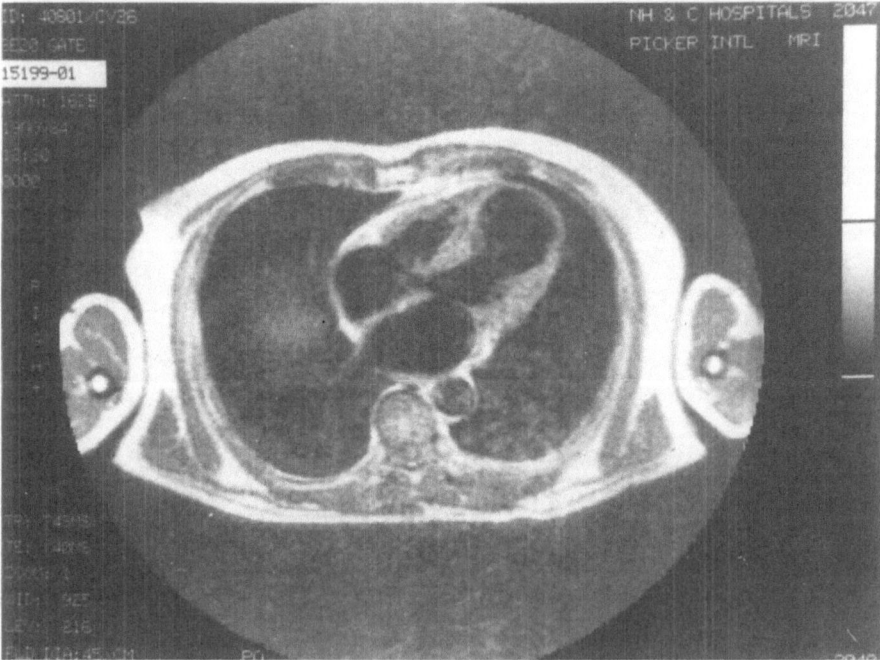

b

Fig. 6.2. Transverse sections at **a** end-diastole and **b** end-systole in a patient with previous myocardial infarction and a left ventricular aneurysm. The basal myocardium contracts but the apex is thin and dyskinetic.

Magnetic resonance may be used in the future to produce such parametric images showing not only wall motion and thickness, but also other parameters such as rates and timing of contraction. Timing is an important feature of regional ventricular function and abnormal areas may contract later than normal (Walton et al. 1981). This introduces inaccuracies into measurements made from a single systolic image which are avoided by cine imaging.

Valvular Regurgitation

The left and right ventricular stroke volumes are identical over the period of imaging only if there is no valvular regurgitation or intracardiac shunting. If there is left-sided regurgitation the left to right ventricular stroke volume ratio will rise, and if there is regurgitation on the right it will fall (Underwood et al. 1986b). From the stroke volume ratio, the regurgitant fraction can be calculated, although if there is both left- and right-sided regurgitation, the ratio may be high or low depending upon which is greater. Figure 6.3 shows a patient with Marfan's syndrome and aortic regurgitation. The left ventricle is dilated but it has a normal ejection fraction, and the left to right ventricular stroke volume ratio indicates severe regurgitation with a regurgitant fraction of 70%.

Congenital Heart Disease

Magnetic resonance is particularly valuable in congenital heart disease because of its ability to demonstrate anatomy. Functional measurements may also be made and in atrial septal defect the pulmonary to systemic flow ratio may be calculated from the stroke volumes of each ventricle (Underwood et al. 1986b). In Fig. 6.4 an ostium secundum atrial septal defect is seen and the right heart is dilated because of volume overload. The pulmonary to systemic flow ratio is 2.9.

Relaxation Parameters

Spin-lattice and spin-spin relaxation times depend upon local biochemistry, although many factors are involved and changes can only be interpreted empirically. Both T_1 and T_2 increase within a number of hours of the onset of myocardial ischaemia, in part due to an increased water content (Brown et al. 1985; Johnston et al. 1985; Pflugfelder et al. 1985; Ratner et al. 1985). Acute infarction is seen as an area of increased signal in the T_2 weighted spin echo image and over a period of 3–4 months T_2 values fall again to below those of normal myocardium (McNamara et al. 1985; Higgins et al. 1984; Thompson et al. 1985). Gadolinium-DTPA may be used to enhance contrast between normal and ischaemic myocardium (Goldman et al. 1982; Brady et al. 1982; McNamara et al. 1984; Wesbey et al. 1984), and it is possible that paramagnetic contrast agents may be used in the same way as thallium 201. Whether this will allow the detection of reversible ischaemia in man is not yet known.

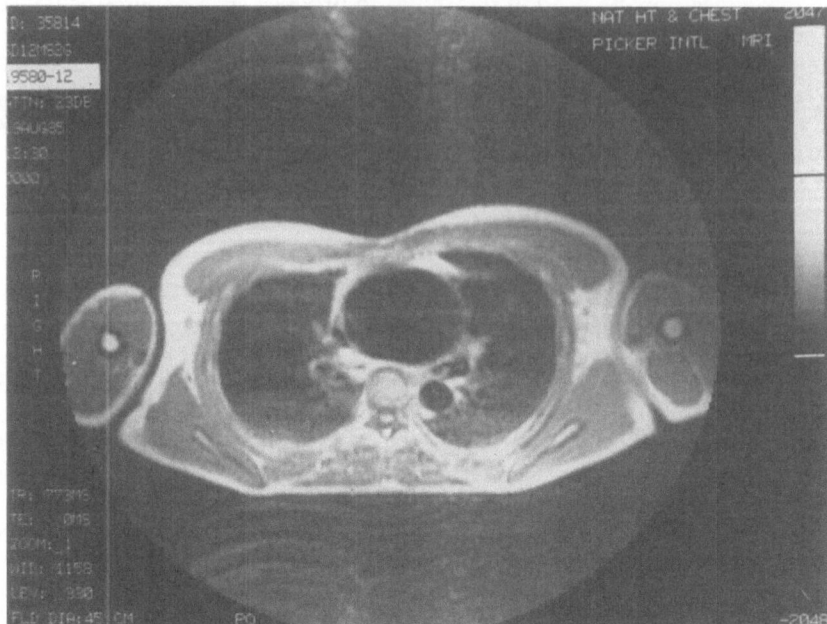

a

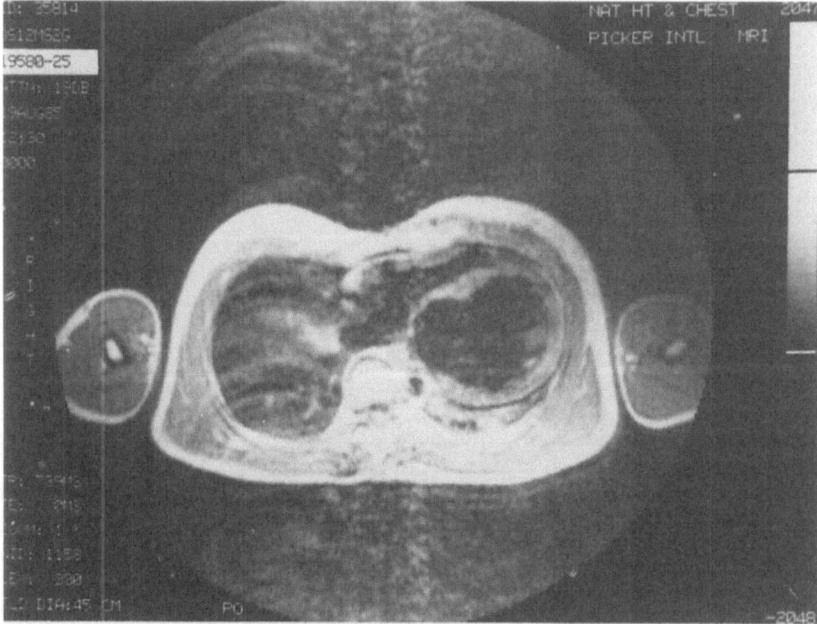

b

Fig. 6.3. Transverse sections **a** through the ascending aorta, and **b** through the left ventricle in a patient with Marfan's syndrome. The ascending aorta is dilated (74 mm diameter) and the heart is pushed into the left chest by the depressed sternum. The left ventricle has a volume of 378 ml with an ejection fraction of 61% and the left to right ventricular stroke volume ratio is 3.4. There is therefore severe left-sided regurgitation (aortic) with a regurgitant fraction of 70%.

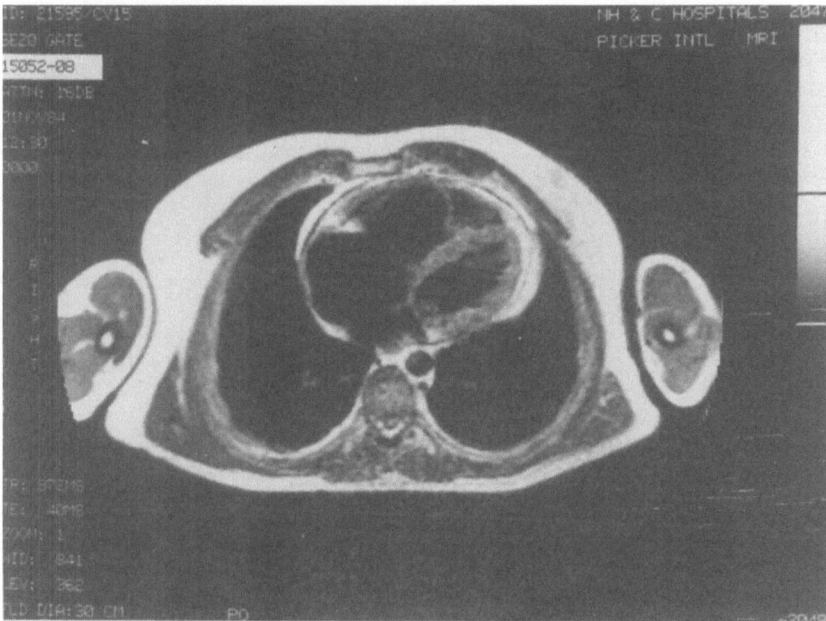

Fig. 6.4. Transverse section showing an ostium secundum atrial septal defect. The left and right ventricular stroke volumes are 69 ml and 200 ml respectively, and the pulmonary to systemic flow ratio is 2.9. Because of the right ventricular volume overload, the right heart is dilated and the moderator band of the right ventricle is well seen.

Magnetic Resonance Angiography

Flowing blood does not give a signal with the spin echo sequences commonly used because it must experience both the 90° and the 180° pulses. If the same section is acquired with a sequence giving signal from moving blood and with one giving no signal, the two images may be subtracted and only the blood will be seen. A simple way of doing this is to use the same sequence at end-diastole and at end-systole, since at end-systole arterial flow will be fast and signal will be lost. The subtracted image will then show only pulsatile arterial flow (Wedeen et al. 1985). This method is, however, sensitive to patient motion between scans or motion of the vessel between diastole and systole and is not suitable for the coronary arteries. A more versatile method is to use two different sequences at the same part of the cardiac cycle and to interleave the sequences in the same acquisition (Nayler et al. 1986). One sequence may be a conventional spin echo sequence, and the other a very short spin echo or field echo sequence which gives signal even from rapidly moving blood. Figure 6.5 shows a magnetic resonance angiogram of the neck vessels acquired in this way. Several sections have been summed to give an image similar to conventional angiography.

Although it is possible to image large vessels in this way, the resolution of most machines is not yet adequate to define stenoses in the coronary circulation. Figures 6.6

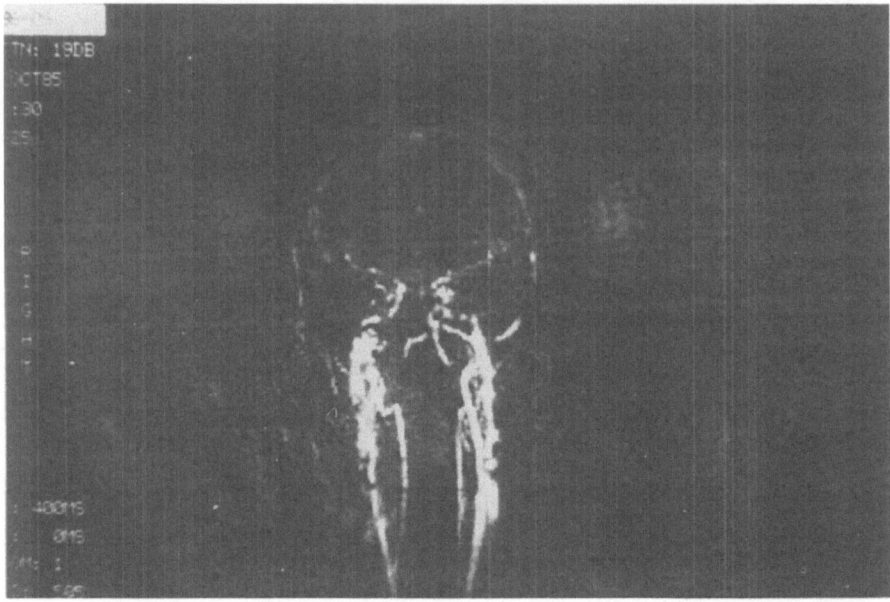

Fig. 6.5. Magnetic resonance angiogram of the vessels of the head and neck. The image is created by subtracting two images, one acquired with a sequence giving signal from flowing blood, and one with a sequence giving no signal from flowing blood. Five contiguous 10-mm sections have been summed.

and 6.7 show the branches of the left coronary artery and a coronary artery bypass graft respectively. When resolution is improved it will be possible to provide non-invasive coronary arteriography.

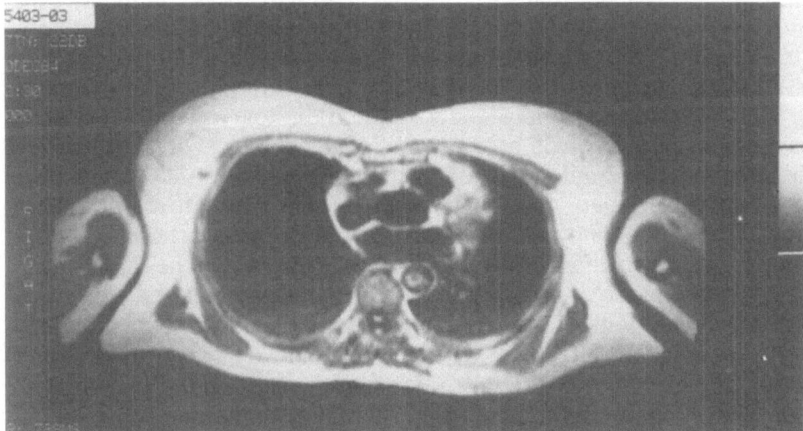

Fig. 6.6. Transverse section through the aortic root showing the left main coronary artery dividing into the anterior descending and the circumflex arteries. The ostium of the right coronary artery is seen anteriorly.

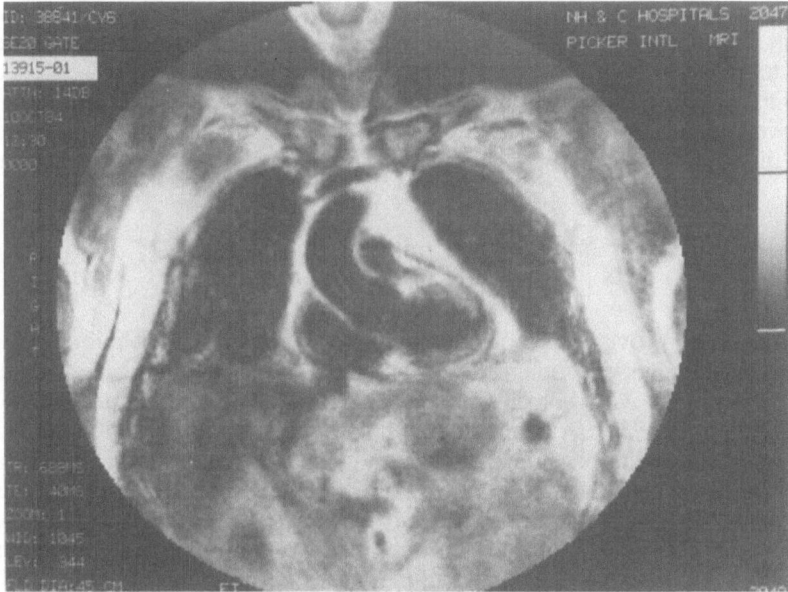

Fig. 6.7. Coronal section showing a bypass graft to the left circumflex coronary artery.

Flow Imaging

Magnetic resonance has been used in industry for the measurement of flow for many years but it is only recently that accurate quantification of blood flow has been possible in vivo. Three main techniques have been developed (Axel 1984).

Flow Between Slices

This technique uses a spin echo sequence but the 90° and 180° pulses are applied in different slices and only blood that has flowed between the two experiences both pulses and gives a signal. The time between the pulses and the distance between the slices dictate the velocity of the blood that is seen, but although this technique is of value in locating blood vessels it has proved difficult to use for quantitative measurements.

Flow into a Saturated Plane

Following a 90° pulse the component of magnetisation in the field direction, M_z, is reduced to zero and if a field gradient is applied in the xy plane, M_{xy} will also be abolished because of rapid de-phasing. There will then be no magnetisation within the plane and it is said to be saturated. If a 90° pulse is applied to a saturated slice, no signal is observed because there is no residual magnetisation but if blood with full magnetisation flows into the slice it will give a strong signal. The intensity will be proportional to the velocity of flow since high velocities will lead to more magnetised blood within the slice. A ceiling will be reached when the blood is fully replaced in the time between saturation and the 90° pulse. In practice this technique has also proved difficult to apply since signal intensity is affected by variables other than velocity.

Phase Mapping

Phase mapping has proved the most accurate of the flow techniques for quantification. The magnetic resonance signal has phase as well as amplitude and the phase may be encoded with velocity information using a combination of field gradients (Bryant et al. 1984). A conventional image is a display of amplitude but a display of phase produces a parametric image showing velocity. An example of a phase map through the great vessels is shown in Fig. 6.8 with blood flowing towards the feet shown in white and towards the head in black. These maps display the velocity of flow quantitatively, and they may show flow within the plane as well as perpendicular to it. Figure 6.9 is a coronal map of the left ventricle and ascending aorta during systole with velocity shown in the z direction (head to feet). The maps may be acquired using a cine sequence, and integration of velocity with time gives bulk flow (Nayler et al. 1985).

Phase mapping is already of value in the investigation of the peripheral arterial system and in patients with valvular regurgitation. High resolution maps of coronary artery flow promise to be of enormous value in the management of coronary artery disease.

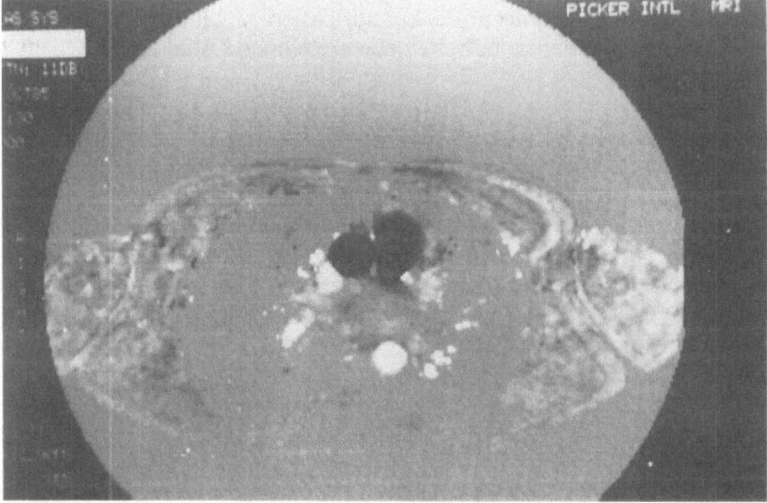

Fig. 6.8. Transverse phase map in mid-systole through the great arteries showing flow perpendicular to the plane of the image. Velocities towards the head are shown in *black* (ascending aorta and pulmonary artery) and towards the feet in *white* (descending aorta and superior vena cava).

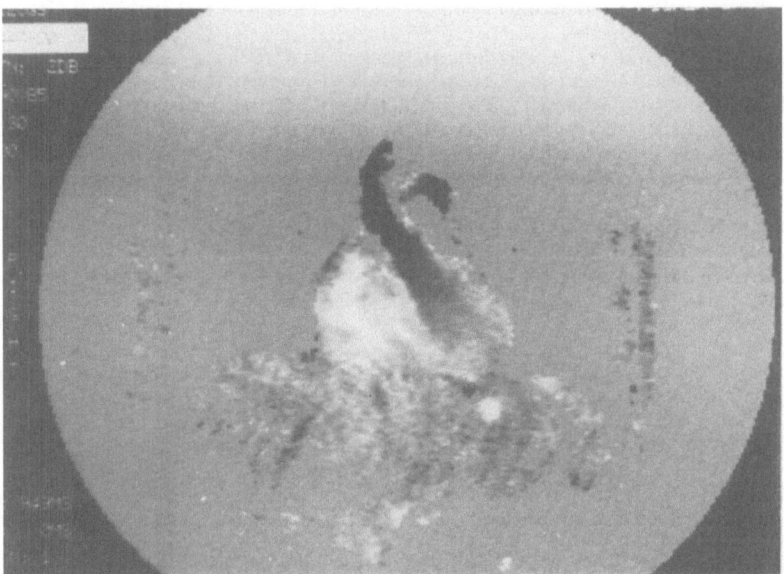

Fig. 6.9. Coronal phase map in mid-systole through the aortic valve showing flow in the plane of the image from head to feet. Flow towards the head is seen in *black* in the ascending aorta, and towards the feet in *white* in the right atrium.

The Future

It is clear that magnetic resonance is particularly suitable for the assessment of cardiac anatomy and ventricular function. Although echocardiography and radionuclide ventriculography provide similar information, magnetic resonance is in its infancy and many future improvements can be expected. For instance, machines will become cheaper and more widely available and scanning will become more rapid. A real time three-dimensional display of the heart is a realistic prospect. Its unique role, however, will be in the study of biochemistry and blood flow. The combination of these studies with non-invasive coronary arteriography will provide magnetic resonance with an important place in the future management of cardiovascular disease.

References

Akins EW, Hill JA, Fitzsimmons JR, Pepine CJ, Williams CM (1985) Importance of imaging plane for magnetic resonance imaging of the normal left ventricle. Am J Cardiol 56: 366–372

Axel L (1984) Blood flow effects in MRI. AJR 143: 1167–1174

Bailes DR, Gilderdale DJ, Bydder GM, Collins AG, Firmin DN (1985) Respiratory ordered phase encoding (ROPE): a method for reducing respiratory motion artefacts in MR imaging. J Comput Assist Tomogr 9: 835–838

Brady TJ, Goldman MR, Pykett IL et al. (1982) Proton nuclear magnetic resonance imaging of regionally ischaemic canine hearts: effect of paramagnetic proton signal enhancement. Radiology 144: 343–347

Brown JJ, Strich G, Higgins CB, Gerber KH, Slutsky RA (1985) Nuclear magnetic resonance analysis of acute myocardial infarction in dogs: the effects of transient coronary ischaemia of varying duration and reperfusion on spin lattice relaxation times. Am Heart J 109: 486–490

Bryant DJ, Payne JA, Firmin DN, Longmore DB (1984) Measurement of flow with NMR imaging using a gradient pulse and phase difference technique. J Comput Assist Tomogr 8: 588–593

Byrd BF III, Schiller NB, Botvinick EH, Higgins CB (1985) Normal cardiac dimensions by magnetic resonance imaging. Am J Cardiol 55: 1440–1442

Feiglin DH, George CR, MacIntyre WJ et al. (1985a) Gated cardiac magnetic resonance structural imaging: optimization by electronic axial rotation. Radiology 154: 129–132

Feiglin DH, Moodie DS, O'Donnell JK, Go RT, Sterba R, MacIntyre WJ (1985b) Evaluation of congenital heart disease by cine magnetic resonance imaging (abstract). Eur J Nucl Med 11: A18

Fisher MR, von Schulthess GK, Higgins CB (1985) Multiphasic cardiac magnetic resonance imaging: normal regional left ventricular wall thickening. AJR 145: 27–30

Friedman BJ, Waters J, Kwan OL, DeMaria AN (1985) Comparison of magnetic resonance imaging and echocardiography in determination of cardiac dimensions in normal subjects. J Am Coll Cardiol 5: 1369–1376

Goldman MR, Brady TJ, Pykett IL et al. (1982) Quantification of experimental myocardial infarction using nuclear magnetic resonance imaging and paramagnetic ion contrast enhancement in excised canine hearts. Circulation 66: 1012–1016

Higgins CB, Lanzer P, Stark O et al. (1984) Imaging by nuclear magnetic resonance in patients with chronic ischaemic heart disease. Circulation 69: 523–531

Higgins CB, Byrd BF, McNamara MT et al. (1985a) Magnetic resonance imaging of the heart: a review of the experience in 172 subjects. Radiology 155: 671–679

Higgins CB, Kaufman L, Crooks LE (1985b) Magnetic resonance of the cardiovascular system. Am Heart J 109: 136–152

Jenkins JPR, Waterton JW, Love HG, Ping ZX, Isherwood I, Rowlands D (1985) Magnetic resonance imaging of the human heart (abstract). Br Heart J 53: 91

Johnston DL, Brady TJ, Ratner AV et al. (1985) Assessment of myocardial ischaemia with proton magnetic resonance: effects of a three hour coronary occlusion with and without reperfusion. Circulation 71: 595–601

Longmore DB, Klipstein RH, Underwood SR et al. (1985) The dimensional accuracy of magnetic resonance in studies of the heart. Lancet I: 1360–1362

McNamara MT, Higgins CB, Ehman RL et al. (1984) Acute myocardial ischaemia: magnetic resonance contrast enhancement with gadolinium-DTPA. Radiology 153: 157–163

McNamara MT, Higgins CB, Schechtmann N et al. (1985) Detection and characterisation of acute myocardial infarction in man with use of gated magnetic resonance. Circulation 71: 717–724

Nayler GL, Firmin DN (1986) Multislice MRI angiography. Proceedings of the 4th annual meeting of the Society for Magnetic Resonance Imaging. Society for Magnetic Resonance Imaging, McLean (in press)

Nayler GL, Firmin DN, Longmore DB, Randell CP (1985) Cine MR blood flow imaging (abstract). Radiology 157(P): 313

Ordidge RJ, Mansfield P, Doyle M, Coupland RE (1982) Real time moving images by NMR. In: Witcofski RL, Karstaedt N, Partain CL (eds) Proceedings of the international symposium in NMR imaging. Bowman Gray School of Medicine Press, Winston Salem NC, pp 89–92

Pflugfelder PW, Wisenberg G, Prato FS, Carroll SE, Turner KL (1985) Early detection of canine myocardial infarction by magnetic resonance imaging in vivo. Circulation 71: 587–594

Ratner AV, Okada RD, Newell JB, Pohost GM (1985) The relationship between proton nuclear magnetic resonance relaxation parameters and myocardial perfusion with acute coronary arterial occlusion and reperfusion. Circulation 71: 823–828

Steiner RE, Bydder GM, Selwyn A et al. (1983) Nuclear magnetic resonance imaging of the heart. Current status and future prospects. Br Heart J 50: 202–208

Thompson RC, Johnston D, Dinsmore RO et al. (1985) Serial magnetic resonance imaging in patients following recent myocardial infarction: serial appearance and characterisation of the healing myocardium (abstract). In: Proceedings of the 4th annual meeting of the Society of Magnetic Resonance in Medicine. Society of Magnetic Resonance in Medicine, Berkeley, pp 678–679

Underwood SR, Walton S, Laming PJ et al. (1985a) Left ventricular volume and ejection fraction determined by gated blood pool emission tomography. Br Heart J 53: 216–222

Underwood SR, Walton S, Laming PJ et al. (1985b) Three dimensional quantification of left ventricular wall motion by ECG gated blood pool emission tomography (abstract). Br Heart J 53: 90

Underwood SR, Rees RSO, Savage PE et al. (1986a) The assessment of regional left ventricular function by magnetic resonance. Br Heart J 54 (in press)

Underwood SR, Klipstein RH, Firmin DN et al. (1986b) Magnetic resonance quantification of atrial shunting and valvular regurgitation (abstract). Br Heart J 54 (in press)

Walton S, Yiannikas J, Jarritt PH, Brown NJG, Swanton RH, Ell PJ (1981) Phasic abnormalities of left ventricular emptying in coronary artery disease. Br Heart J 64: 245–253

Wedeen VJ, Meuli RA, Edelman RR et al. (1985) Projective imaging of pulsatile flow with magnetic resonance. Science 230: 946–948

Wesbey GE, Higgins CB, McNamara MT et al. (1984) Effects of gadolinium-DTPA on the magnetic relaxation times of normal and infarcted myocardium. Radiology 153: 165–169

7. Paramagnetic Pharmaceuticals for Functional Studies

L.D. Hall and P.G. Hogan

Introduction

Magnetic resonance imaging (MRI) has advanced technically to the point where it is already regarded by many radiologists as the preferred imaging method for most studies of the brain. However, there is less uniform opinion concerning the potential of MRI in a number of other important clinical areas, e.g. in studies of the abdomen where CT X-ray techniques are already well established and where the effects of motion, both voluntary and peristaltic, serve to compound the already major problems posed for MRI by the poor intrinsic tissue contrast. Furthermore, although MRI methods have already made spectacular advances for detecting aberrant pathology since the first speculative suggestion of Damadian (perhaps best epitomised by the first results on multiple sclerosis from the Hammersmith team), even in the brain unequivocal identification of the precise pathology of a lesion by MRI cannot be guaranteed, e.g. distinction between a tumour, and the surrounding tissue and oedema.

It has been suggested that these and other limitations of the MRI method may be minimised by the use of "image contrast" agents. These are exogenous chemicals administered to the patient which, by influencing the magnetic resonance properties of the water in the region of the pathology, serve to heighten the contrast between that tissue and its surroundings.

On the basis of the references in this article and the already substantial body of knowledge concerning X-ray contrast agents, the characteristic properties of an "ideal" contrast agent for MRI are obvious. Fundamentally, it should be non-toxic and provide effective contrast when used in very small quantities; in turn, these characteristics imply that it should have an appropriate balance of pharmaceutical properties such as tissue uptake and excretion rate; ideally, the material should be organ specific and, in a perfect world, be specific to just one particular pathology.

The efficacy of an MR image contrast agent depends critically on the interplay between three distinct concepts: (a) that of the meaning of *intensity* in an MR image, (b) that of the influence of *nuclear magnetic relaxation* on those measured intensities and (c) that of the effect of *paramagnetic chemicals* on the relaxation rates of the nuclear spins (principally the water protons) of the tissues which are under investigation.

Most MR images are displayed on a grey scale with white implying the highest "effective intensity" and black the lowest. This *effective intensity* (EI) = (concentration) = (My), where (concentration) is the concentration of the magnetic nuclides in the tissue and (My) is the amount of their magnetisation present at the actual time of the MR measurement. Although there is intrinsic variation of water concentration between different tissues, it is the substantial tissue dependence of (My) which provides the major source of image intensity variations and hence image contrast.

This relationship stems from the fact that each magnetic nucleus has two characteristic relaxation times, T_1 and T_2, which govern the rate at which the magnetisation of that nuclear species can be accurately measured. In what follows we shall emphasise effect on T_1 values and not mention T_2. Consider then, an MR imaging protocol in which the magnetisation (My) is measured repeatedly at a predetermined rate. Tissues which have a short T_1 value will recover all their (My) between successive scans and hence their perceived intensity in the final image will accurately reflect their water content; that intensity will be high and hence the tissue will show, for example, as white. On the other hand, tissues with T_1 values which are long compared with the experimental repeat time will progressively lose magnetisation and will show as a darker colour on the image intensity grey scale.

It is this variation of endogenous tissue relaxation characteristics which provides the wealth of image contrast which makes MRI such a powerful diagnostic procedure. This is especially true since these can be further exaggerated by intelligent manipulation of the nuclear magnetisation, as in the well known inversion recovery for T_1 variations, or the spin echo sequence for T_2 variations. Unfortunately, there are some regions of the body where this endogenous contrast is insufficient for adequate tissue contrast and it is for such regions that the use of exogenous image contrast agents has been suggested.

These agents are chosen because they can alter the relaxation times of protons in those tissues with which they are in contact. Thus, if an agent which can shorten the T_1 values is introduced into a particular organ then the perceived intensity of that particular region of the final image can be enhanced by choice of suitable imaging times. Obviously the greater the specificity with which an agent can be delivered to a specific tissue, the greater will be the induced contrast.

The major mechanism whereby a paramagnetic contrast agent induces its effect, the dipole–dipole mechanism, involves a through space interaction between the magnetic field generated by the rapidly moving paramagnetic centre and the precessing nuclear spin which is being studied. Although the efficiency with which these two entities communicate can be described explicitly, the relationship is not simple, even for isotropic media, and for human tissues the situation is almost certainly not quantifiable. It is obvious that the relaxation efficiency of a particular paramagnetic species will be enhanced if there are more unpaired electrons; thus one might expect that a metal with several unpaired electrons would be better than one which has only one. It is also intuitively reasonable to expect that the more water molecules which come within the sphere of paramagnetic influence per unit of time, the greater will be the perceived overall effect; thus, a free metal in water should be more accessible than one which is chelated in a more hydrophobic pocket. Unfortunately, a number of more subtle factors also come into play, including the relaxation times of the unpaired electrons and also the overall rate at which the paramagnetic ensemble is tumbling in solution. It is for that reason that insight into the efficacy of image contrast agents can only come from an experimental approach.

At present the most widely used agent is gadolinium-DTPA (Gd-DTPA). This appears to have many desirable features and its development provides a textbook example for the early stages of any future development.

All compounds used so far (Wolf et al. 1985) can be subdivided into one of the following categories:

1. Paramagnetic metal species
2. Ferromagnetic metal species
3. Stable free radicals
4. Oxygen carriers
5. Susceptibility agents
6. Density substitution agents

In what follows we summarise briefly these chemical substances and their reported uses.

Paramagnetic Metal Species

In 1976 (Frank et al. 1976) demonstrated in vitro how proton NMR imaging can highlight induced canine myocardial infarction and pulmonary extravascular water. In 1978 Lauterbur et al. (1978a, b) reported how infusing the heart with a saline solution of $MnCl_2$ can significantly contrast enhance the induced myocardial infarction. More recently many manganese salts and complexes have been successfully employed as NMR image contrast agents (Koutcher et al. 1984; Hansen et al. 1980; Engelstad and Brasch 1984; Schumacher et al. 1985). There have also been studies on many other paramagnetic transition and rare earth metals. Ferric and ferrous salts have been used as oral and rectal contrast agents (Wesbey et al. 1983a, b, 1984a) and some have been intravenously administered (Chen et al. 1984). Copper salts and complexes have been studied, and there has been considerable progress with the complexes of chromium, gadolinium and other rare earth metals (Eu, Dy, Pr etc.) (Runge et al. 1983a, c, 1984a–c, 1985a; Carr 1984; Carr et al. 1984a, b; Unger et al. 1985).

Classes of Paramagnetic Complexes

Briefly these can be subdivided as follows:

Oral agents
Soluble metal ions: Cr^{3+}, Cu^2, Fe^{3+}, Mn^{2+}

Soluble metal ion complexes:
  chromium ethylenediaminetetraacetic acid, Cr-EDTA (Fig. 7.1)
  ferric ammonium citrate (Fig. 7.2)

Insoluble particulate species:
  gadolinium oxalate, $Gd_2(C_2O_4)_3$ (Fig. 7.3)
  Cr trisacetylacetonate (Fig. 7.4)

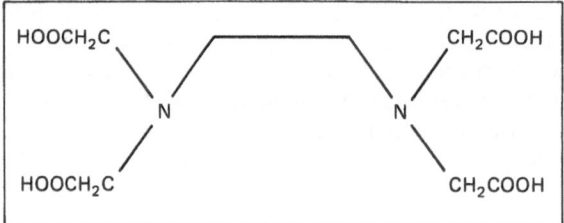

Fig. 7.1.
Ethylenediaminetetraacetic acid
(EDTA).

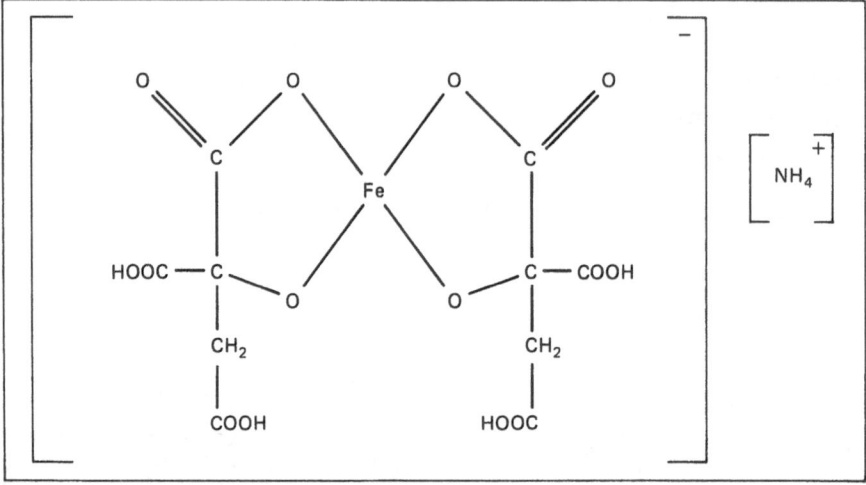

Fig. 7.2. Ferric ammonium citrate.

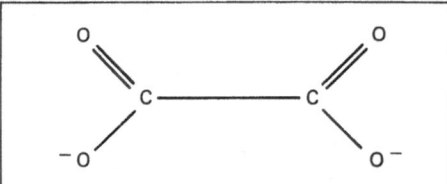

Fig. 7.3. Oxalate ($C_2O_4{}^{2-}$).

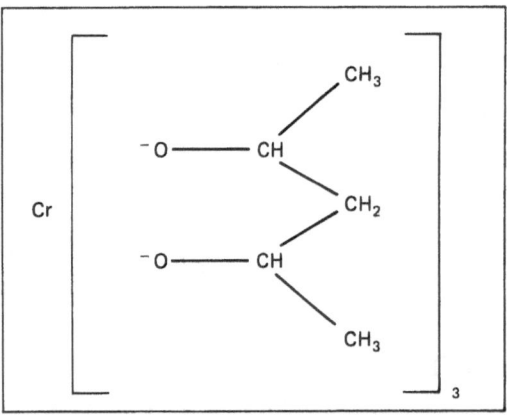

Fig. 7.4. Chromium
trisacetylacetonate.

Intravenous agents

Metal ion chelates/complexes:
  gadolinium diethylenetriamine pentacetic acid, Gd-DPTA (Fig. 7.5)
  Cr-EDTA, Fe-EDTA
Metalloporphyrins:
  Tetraphenylsulphonyl porphyrins, TPPS$_4$ (Pd, Cu, Mn, Fe) (Fig. 7.6)
  Mn hematoporphyrin derivative, HPD (Fig. 7.7)
  Mn protoporphyrin IX (Fig. 7.8)

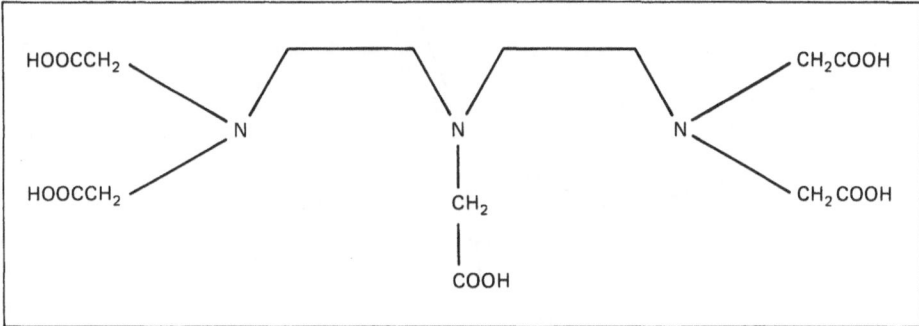

Fig. 7.5.   Diethylenetriamine pentaacetic acid (DTPA).

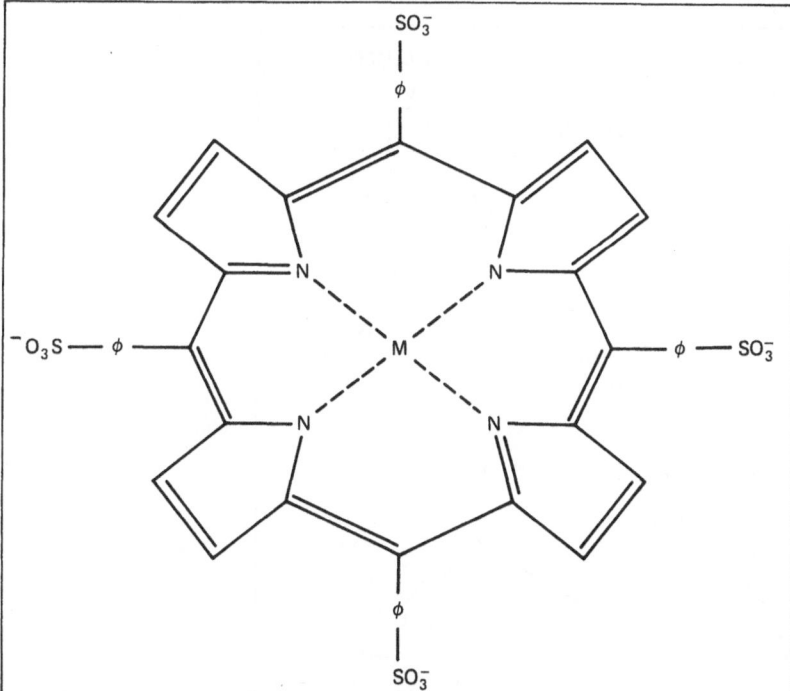

Fig. 7.6.   Tetraphenylsulphonyl porphyrin (TPPS$_4$).

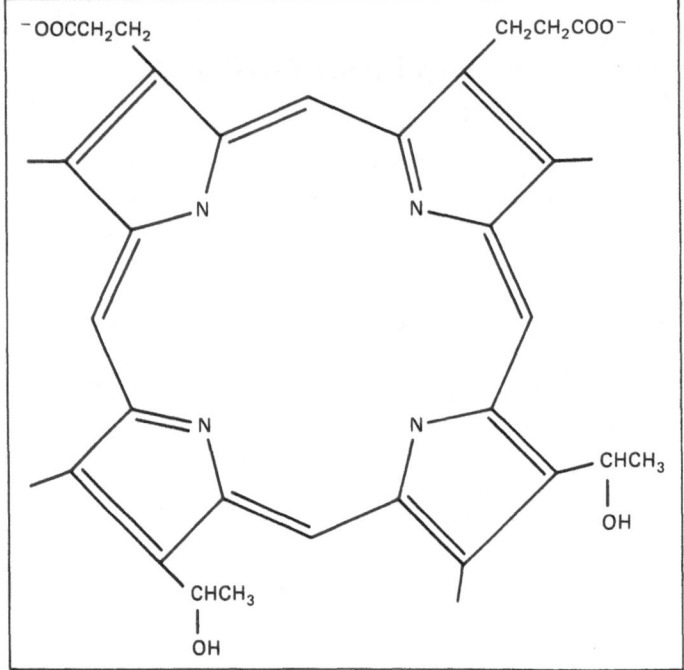

Fig. 7.7. Hematoporphyrin (HPD).

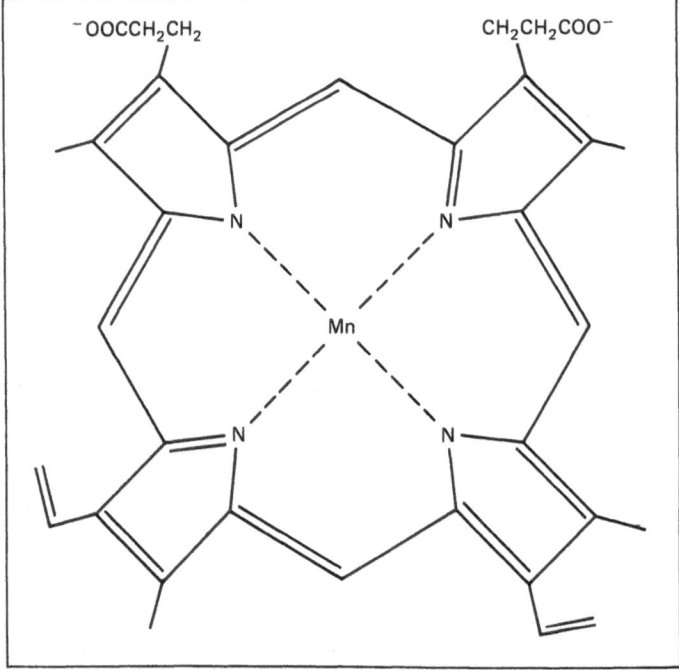

Fig. 7.8. Manganese protoporphyrin IX.

We now summarise these species into their specific clinical applications:

Heart Imaging

There has been considerable use of manganese (II) salts to successfully contrast enhance infarcted areas of the myocardium (Goldman et al. 1982a; Alfidi et al. 1982; Mendonca-Dinas et al. 1983; Jackson et al. 1985; Gore et al. 1983; Runge et al. 1983a; Goldman et al. 1982b; Brady et al. 1982; Brasch 1983; Pohost et al. 1982; Hollis et al. 1979) due to its attractive properties of a relatively high magnetic moment and well established blood kinetics (Weinmann 1984; Pople et al. 1959). Accurate determination of the infarct size provides prognostic information on patients with coronary disease and allows the progress of therapy to be monitored. Manganese is actively taken up by viable myocardial cells and has a short half-life in the blood pool, providing a high manganese to blood ratio (Wolf and Baum 1983; Budinger 1982).

The monoclonal antibody to cardiac myosin, which specifically localises in infarcted myocardium, has been labelled with Mn-DPTA (Brady et al. 1983; Lauffer 1984). This species acts as a tissue specific contrast agent.

Gadolinium has been used in the complex form, Gd-DTPA, to enhance acute myocardial ischemia (Carr et al. 1984b; McNamara et al. 1984) Wesbey et al. 1984a, b; Fobben and Wolf 1983) and infarcted and normal myocardium (Englestad and Brasch 1984b; Fobben and Wolf 1983).

Gastrointestinal Tract Imaging

Several iron salts have been orally or rectally administered to contrast enhance the GI tract: ferric chloride (Wesbey et al. 1983a; Young et al. 1981), ferric ammonium citrate (Wesbey et al. 1983b; Young et al. 1982), ferrous sulphate heptahydrate and iron dextran (Alfidi et al. 1982; Runge et al. 1983b).

Both ferric ammonium citrate (Geritol) and ferrous sulphate heptahydrate (Fer-In-Sol) shorten T_1 and T_2 of the water protons in the contrast media filled GI tract, permitting clear distinction between GI structures and adjacent abdominal viscera. Geritol in particular provided clear identification of the gastric lumen and the gastric wall (Stark et al. 1984).

Ferric chloride, $FeCl_3.6H_2O$, provides enhancement of the fundus of the stomach but causes GI irritation (Spector 1956), and it has been found (Young et al. 1981, 1982) that the ferric iron has seven times less adverse effects than the ferrous iron.

Cr-EDTA taken orally has also provided contrast enhancement of the GI tract (Runge et al. 1983a, c). It is a stable chelate, poorly absorbed by the GI tract and the portion that is absorbed is readily excreted by glomerular filtration leading to minimal toxicity.

Insoluble particulate paramagnetic species such as chromium tris acetylacetonate and gadolinium oxalate are inert and pass through the GI tract intact with no absorption, yet still provide significant contrast enhancement (Gore et al. 1983; Runge et al. 1983a–c, 1985b; Burnett et al. 1985). They have been used to identify the pancreatic head by opacification of the C-loop of the duodenum, distinguishing soft tissue masses in the abdomen from fluid- and faeces-filled bowel loops.

Urographic and Renal Imaging

Cr-EDTA and Gd-DTPA are used as flow agents and as general organ contrast agents enhancing the spleen, liver and kidneys in particular (Engelstad and Brasch 1984; Runge 1983a, c, 1984a; Burnett et al. 1985; Clanton et al. 1984). These agents are administered intravenously: there is selective accumulation of the agent by the kidneys followed by excretion and they thus yield useful information about renal function and urological anatomy (Brasch 1983). The chelates still remain stable after intravenous injection.

Such contrast enhancement enables abnormalities to be both identified and quantified where non-enhanced NMR imaging fails to show any differences. For example, London et al. (1983) produced renal ischaemia in a rat by surgically ligating the left renal artery but NMR scans 4 h after surgery could not show any difference between the normal right kidney and the ischaemic left kidney.

These agents are excreted by glomerular filtration (Runge et al. 1983c; Stacy and Thorburn 1966) and through this means, Gd-DTPA has been used to contrast enhance acute renal hydronephrosis caused by a surgically obstructed right ureter (Runge et al. 1983c), a parapelvic kidney cyst (Schorner et al. 1984a, b) and normal and abnormal kidney (Weinmann 1984; Wolf and Fobben 1984; Brasch et al. 1984). Recent studies with polyazamacrocyclic ligands (Geraldes et al. 1985; Josipowicz et al. 1985; Knop et al. 1985) show Gd-DOTA (1,4,7,10-tetraazacyclododecane-N,N', N'',N'''-tetraacetic acid) (Fig. 7.9) in particular to have a higher thermodynamic and kinetic stability compared to Gd-DTPA, yet it has similar pharmacokinetics and thus has the potential to be a better contrast agent for kidneys and liver.

Gd-DTPA has no specific organ uptake and so provides relatively inefficient imaging of the liver and spleen because of the low concentrations accumulated in these organs after administration. However, it is possible to link gadolinium species to liposomes (Buonocore et al. 1985), and by varying the size of the lipid encapsule it is possible to transport paramagnetic ions to individual organs in varying but predictable concentrations. The T_1 of both liver and spleen in mice has been lowered by 25%-35% compared to controls after intravenous administration.

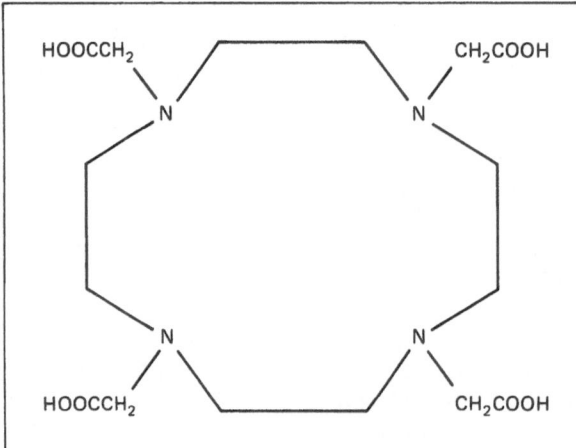

Fig. 7.9   1,4,7,10-tetraazacyclododecane-N,N',N'',N''' -tetraacetic acid.

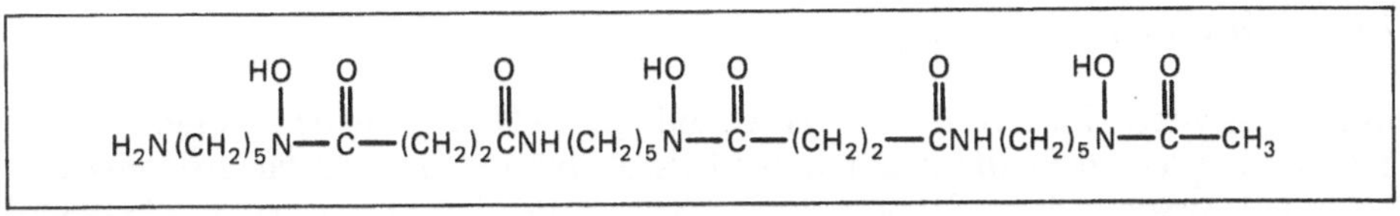

Fig. 7.10. Desferrioxamine B.

Colloidal manganese sulphide (Chilton et al. 1984) has been used to study the reticulo-endothelial system (RES) by significantly altering the T_1 of the liver and kidneys (Wolf and Baum 1983; Budinger 1982); the colloidal species has quite high resistance to in vivo metabolism. Manganese protoporphyrin IX has shown potential usefulness as an intravenous contrast agent for the kidneys and the liver in particular (Jackson et al. 1985).

Many iron binding ligands — catecholates, hydroxamates, amino-carboxylic acids (EDTA, DTPA), porphyrins and oxines — have demonstrated contrast properties (Chen et al. 1984; Carr et al. 1984a) primarily in the kidney. In vivo urographic enhancement is possible with the hexadentate siderphore desferrioxamine B (Fig. 7.10) (White et al. 1985), which has enhanced hydronephrotic kidney and the liver. The ferric chelate of 2,3-dihydroxybenzoic acid (2,3-DHB) (Fig. 7.11) is an effective urographic contrast agent (Engelstad and Brasch 1984) with low toxicity and high stability. The complex iron (III) EHGP (ethylene bis-2-hydroxyphenylglycine (Fig. 7.12), is a hepatobiliary contrast enhancer producing up to seven times greater increase in signal intensity in the liver compared to Fe-DTPA (Lauffer et al. 1985a, b; Grief et al. 1985). Its prolonged but efficient clearance enables facile imaging at non-toxic doses.

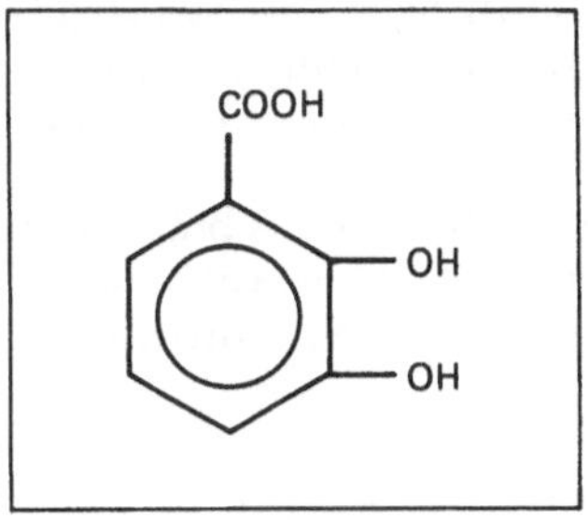

Fig. 7.11. 2,3-Dihydroxybenzoic acid (2,3-DHB).

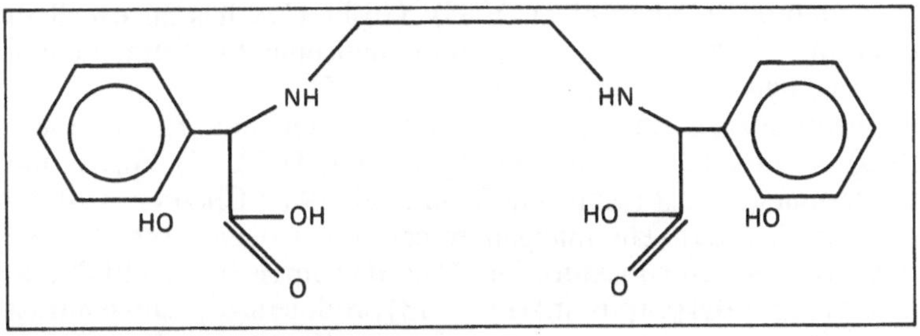

Fig. 7.12. Ethylene bis-2-hydroxyphenylglycine.

Tissue and Vascular Imaging

Intravenously administered Cr-EDTA and Gd-DTPA provide strong contrast enhancement to aid in the evaluation of tissue perfusion, vascular anatomy and isomagnetic tissue lesions (Runge et al. 1983a, c; 1984b, c; Goldman et al. 1982b; Weinmann et al. 1984; Fobben and Wolf 1983; Strich et al. 1985).

After intravenous injection, Gd-DTPA is known to distribute primarily in the intravascular, extracellular space and quickly diffuses from this plasma into the extravascular, interstitial space and so enhances tissue with a high proportion of interstitial fluid, e.g. oedematous tissue.

The metalloporphyrin manganese tetraphenylsulphonylporphyrin (MnTPPS$_4$) has demonstrated improved NMR image visualisation of the vascular wall by altering the proton relaxation times of plaque tissue, in which the porphyrin specifically localises. Studies were performed (Sohn et al. 1985) in rabbit models with plaques induced either by hypercholesterolaemia or by a chronic indwelling catheter in the aorta and also in postmortem human atheromatous lesions.

Brain Imaging

Functional abnormality of the central nervous system (CNS) where the blood–brain barrier (BBB) is disrupted can be highlighted by contrast agents that pass into cerebral tissue only where the BBB has broken down. The BBB prevents diffusion of Cr-EDTA and Gd-DTPA into brain tissue, thus having little effect on normal brain tissue but successfully discriminating cerebral oedema from neoplastic diseased tissue (Runge et al. 1984c).

Gd-DTPA has a high neural tolerance coupled with a strong proton relaxation effect (Weinmann et al. 1984) which has made it an ideal agent for selectively enhancing such abnormalities as brain abscesses and cerebral tumours (Carr et al. 1984a; Carr and Gadian 1985; Grossman et al. 1984; Brasch et al. 1984; Felix et al. 1985).

Desferrioxamine B can produce focal T_2 shortening attributable to BBB breakdown in a rat model of radiation cerebritis (Engelstad and Brasch 1984). However, concern over its toxicity limits its administration to lower dose applications such as urography.

Tumour Imaging

Gd-DTPA has been used to contrast enhance cerebral tumours (Grossman et al. 1984) and vaginal carcinoma (Gadian et al. 1985). Gd-DOTA has an excellent tumour-oedema localising effect enabling successful enhancement of human colon carcinoma implanted in athymic nude mice (Knop et al. 1985).

There have also been studies on tumour-specific (Zanelli and Kaelin 1981; Hambright et al. 1975), water-soluble metalloporphyrins [Pd(II), Cu(II), Fe(III) and Mn(II)-TPPS$_4$] as potential contrast agents (Jackson et al. 1985; Chen et al. 1984; Lyon et al. 1985; Fiel et al. 1985). The manganese centre has been shown to give greatest proton relaxation per unit concentration. Mn(III) complexes of TPPS$_4$ and TMPyP (tetra-N-methyl-4-pyridyl porphyrin) (Fig. 7.13) produce significant contrast enhancement in images of human colon carcinoma in mice (Lyon et al. 1985; Knop et al. 1984). Mn(II) labelled hematoporphyrin derivative (HPD), a human tumour

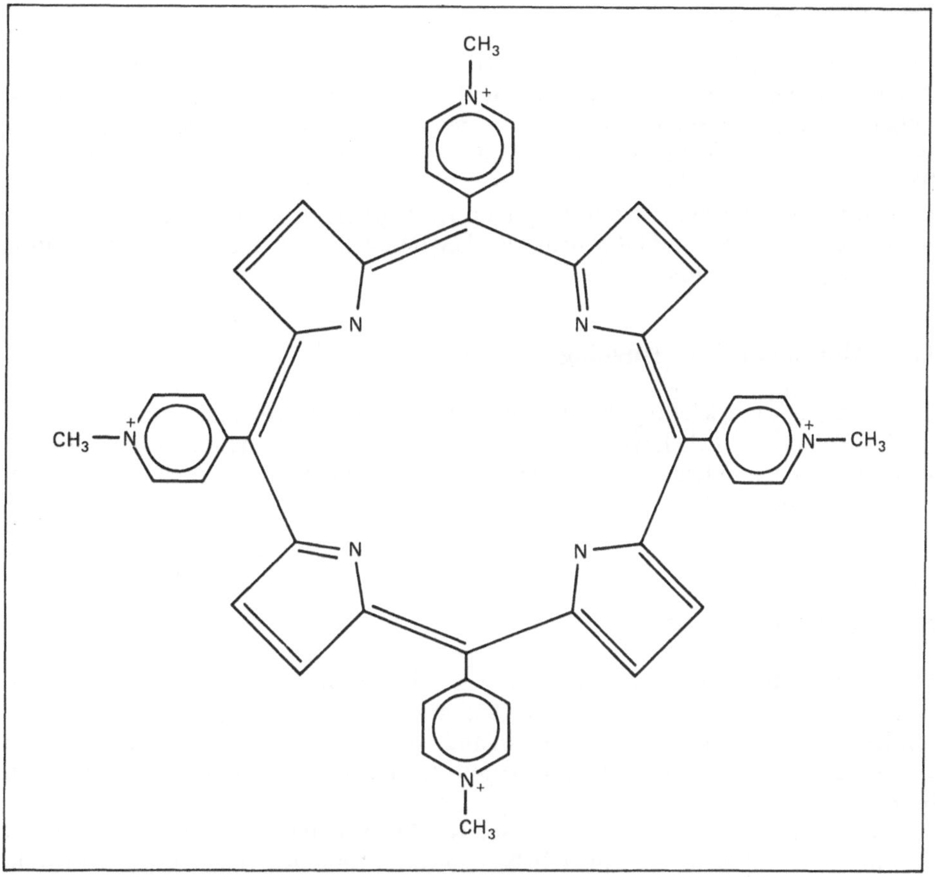

Fig. 7.13. Tetra-*N*-methyl-4-pyridyl porphyrin (TMPyP).

localising agent used for laser phototherapy (Letokhov 1985), shows similar tumour enhancement (Patronas et al. 1985).

Ferromagnetic Metal Species

The contrast effect of paramagnetic compounds is mainly through the lowering of proton T_1 although larger concentrations will reduce T_2. Conversely, ferromagnetic compounds have a stronger relaxation effect on proton T_2. Ferromagnetic foci induce in their vicinity fluctuations in the strength of local magnetic fields and thus affect the NMR signal intensity of surrounding material; such species may thus act as negative NMRI contrast agents (Ohgushi et al. 1978).

RES Imaging

Upon intravasal infusion albumin microspheres containing magnetite (Fe_3O_4) of 2.0 μm average diameter are trapped by the RES and successfully provide contrast enhancement, in particular to hepatic cancer in rats and dogs due to their strong tissue selectivity (Olsson et al. 1985; Saini et al. 1985; Mendonca-Dias et al. 1985). The magnetic microspheres may be targeted to organs close to the patient's surface by controlled, external, strong magnetic field gradients; this provides a possible means to both target and label drugs for NMRI evaluation of blood biodistribution and therapeutic progress.

Gastrointestinal Tract Imaging

Ferromagnetic particles of $0.8-3.0 \times 10^{-6}$ m average diameter and containing 5.0%–32.9% iron per particle have been proposed as oral contrast media for the whole GI tract (Jacobsen and Klaveness 1985) and are complemented by lack of absorption or any systemic toxicity.

Stable Free Radicals

The nitroxide free radical owes its paramagnetic behaviour to the unpaired electron that exists along the N-O bond. By 1965 stable nitroxide free radicals had been synthesised (Stone et al. 1965; Hoffman and Henderson 1961) in which the electron was protected from pairing and so could persist for many hours (or longer) in biological systems.

The nitroxide stable free radicals (NSFR) have tremendous chemical versatility; they have been attached as "spin labels" to a variety of biochemical substrates (Mela and Chance 1965; Dreyer et al. 1983; Kosman et al. 1969; Hsia and Piette 1969; Berliner and Conery 1983) and have more recently been employed in vivo as NMR image contrast enhancers (Mendonca-Dias et al. 1983; Brasch 1983; Brasch et al. 1982a, b, 1983a–d; Koutcher et al. 1984; Runge et al. 1983c; Keana and van Nice 1984; Swartz et al. 1986).

Classes of NSFR

There are many paramagnetic NSFR compounds but there are two major subgroups favoured for NMR image contrast agents:

pyrrolidine-N-oxyl (Fig. 7.14)
piperidine-N-oxyl (Fig. 7.15)

Various substitutions at the R group enable selective tissue targeting of the NSFR to suit each particular medium.

Urographic and Renal Imaging

One commonly used NSFR is "TES" (N-succinyl-4-amino-2,2,6,6,-tetra-methylpiperidine-1-oxyl) (Fig. 7.16) (Engelstad and Brasch 1984; Brasch et al.

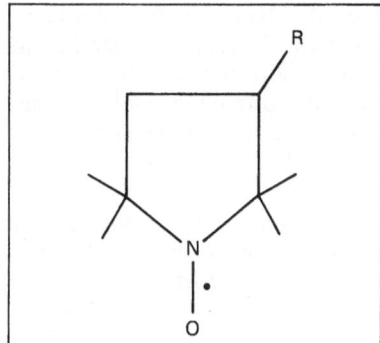

Fig. 7.14. Pyrrolidine-*N*-oxyl.

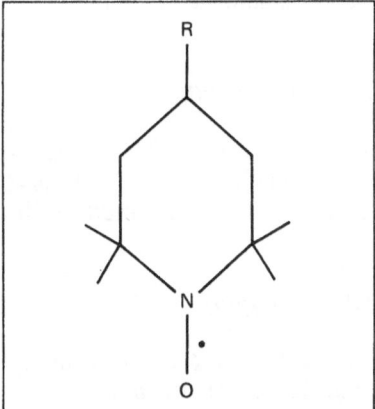

Fig. 7.15. Piperidine-*N*-oxyl.

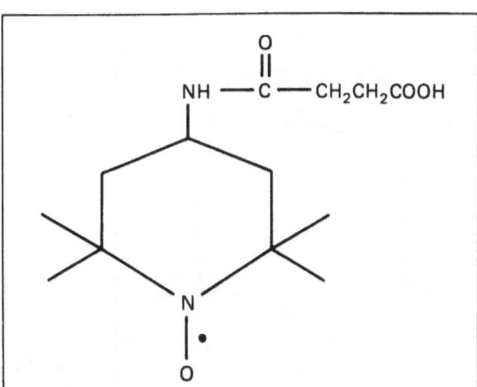

Fig. 7.16 *N*-succinyl-4-amino-2,2,6,6,
-tetramethyl piperidine-1-oxyl.

1983b), which is water soluble and is rapidly excreted in the urine after intravenous injection. It has been tested in animals with experimentally induced unilateral renal ischaemia, renal vascular congestion and hydronephrosis and successfully enhanced these abnormalities, which are not normally distinguished by non-enhanced images. TES has a half-life of approximately 38 min as derived from plasma concentration time and excretion rate time data. TES suffers in vivo metabolism (typical recovery may be only 60% of the initial dose) by reduction to non-paramagnetic hydroxylamines by enzyme systems and antioxidants (Perkins et al. 1980). Its molar relaxativity is lower than certain metal ions (Gd^{3+}, Mn^{2+}, etc.) necessitating higher doses when, unfortunately, toxicology data are sparse (Griffeth et al. 1983).

Brain Imaging

TES has been shown to cross the BBB only at sites of damage and hence provides contrast enhancement of cerebral abnormalities (Engelstad and Brasch 1984; Brasch et al. 1983c; Bydder et al. 1982) such as are induced by bacterial cerebritis and radiation damage in animals.

Tumour Imaging

Another NSFR is "TPC" (Brasch et al. 1983d; McNamara et al. 1985) (a pyrrolidine-N-oxyl where R in Fig. 7.14 is $-CO_2H$), which has been used to contrast enhance tumours in a nude rat model of human renal cell carcinoma.

Tissue Imaging

The NSFR "CAT" (4-trimethylammonium-2,2,6,6-tetramethylpiperidine-1-oxyl) (Fig. 7.17) has been encapsulated in liposomes and used to enhance proton relaxation in macrophages, tissues rich in phagocytic cells (Chan et al. 1985). Peritoneal macrophages collected from BDF_1 mice were incubated with liposomes for periods up to 8 h; progressively larger and more significant amounts of nitroxide-

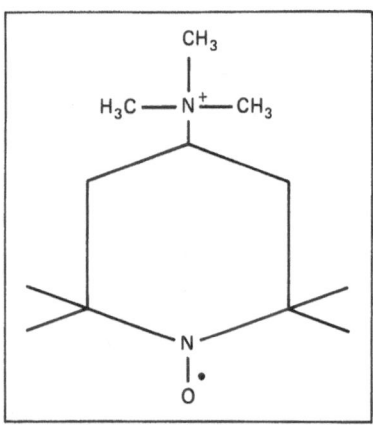

Fig. 7.17.   CAT: 4-trimethylammonium-2,2,6,6,-tetramethylpiperidine-1-oxyl.

laden liposomes were incorporated into the interior of the macrophages. The nitroxide concentration responded to the metabolism of the macrophages and thus has potential as a form of contrast agent that reflects the metabolic state of cells and tissues.

Oxygen and Oxygen Carriers

Molecular oxygen is a paramagnetic gas by virtue of its two unpaired electrons. Oxygen is present in all tissue at an arterial pressure of 100 mm Hg [approximately 3 ml of oxygen dissolved in each litre of blood; an effective concentration of 6×10^6 molecules of oxygen per ml (Mendonca-Dias et al. 1983)]. It has been proven that in water the relaxation rate change is proportional to the oxygen concentration (Gore et al. 1983), but there is only enough paramagnetic oxygen in tissues to contribute to 1%-2% of the total relaxation time. However, there are several ways to oxygen-enrich the bloodstream and thus provide in vivo NMR contrast enhancement.

Classes of Oxygen Carrier

Oxygen is disadvantaged by loss of its paramagnetism on metabolism in vivo (e.g. in the formation of diamagnetic oxyhaemoglobin) and on being chemically complexed to carrier molecules. As a gas it is difficult to direct within the body. There is, however, considerable potential for perfluorinated blood and plasma substitutes, which have excellent oxygen solubility and documented toxicology, to act as NMR image contrast agents through increasing the oxygen concentration in organs of specific interest. In particular, myocardial infarction, the RES, liver, spleen and brain tumour imaging could all be further enhanced (Baldwin and Gill 1982; Faithful et al. 1984; Menasche et al. 1984; Patronas et al. 1983; Mattrey et al. 1982; Bernadov et al. 1985).

Studies on normal saline solution and the artificial blood replacement, FC-43 emulsion (perfluorotributylamine-"Fluosol" (Fig. 7.18), show that increasing the oxygen saturation in solution causes an appreciable decrease in T_1 (i.e. 1 s decrease in a 600 torr pressure increase of oxygen) (Runge 1983a).

Heart Imaging

By inhaling 100% pure oxygen (at 10 ml per min for 10 min), human volunteers have demonstrated an increase in the NMR signal from blood within the left ventricle (oxygenated blood) compared to the right ventricle (deoxygenated blood) (Alfidi et

$$\left[CF_3(CF_2)_3 \right]_3 N$$

Fig. 7.18. Perfluorotributylamine (*F*-tri-*N*-butylamine).

al. 1982; Gore et al. 1983; Runge et al. 1984c; Brasch 1983; Koutcher et al. 1984) and also an increase in the signal density of the intraventricular septum.

Thorax Imaging

Studies with rabbits breathing 100% pure oxygen have shown enhanced images of the thoracic region.

Magnetic Susceptibility Agents

The determination of magnetic susceptibility by NMR has several clinical uses, e.g. the non-invasive measurement of human hepatic iron stores (Brittenham et al. 1982), and there have been studies on the influence of magnetic susceptibility on in vivo proton chemical shift (Pykett and Rosen 1983). If tissue specific agents can be administered which have a high, localised magnetic susceptibility then they will function as another source of NMR contrast agent.

Density Substitution Agents

It is possible to effect contrast enhancement in vivo by altering the water content or by replacing water with another compound of different proton density and relaxation time (Alfidi et al. 1982; Koutcher et al. 1984; Hansen et al. 1980; Lauffer et al. 1985a).

Lipids/Liposomes

It is possible to introduce lipids into cavities like the intestinal tract with enhanced NMR resolution, due to both the lower relaxation time of lipids and their higher volume proton density (Pykett and Rosen 1983; Beall 1982; Newhouse et al. 1982). Oral administration of mineral oil has been shown to opacify bowel loops (Newhouse et al. 1981).

Liposomes are rounded fatty vesicles composed of one or more layers of lipid membrane, generated in varying sizes, and thereby targeted to specific cells such as the reticuloendothelial cells of the liver, spleen and bone marrow (Papahadjopopulos 1978; Sharma et al. 1977). As they are almost entirely composed of lipid they make for ideal contrast enhancers.

Water Content

Changes in tissue hydration can be achieved by administration of diuretics, hormones which can change water content in some tissues (Brasch 1983; Beall 1982), parenteral hypertonic agents and radiographic contrast media. The latter's potential as an NMR contrast agent is due to their hypertonicity; such media become concentrated in the kidney after intravenous injection and cause diuresis; the resulting shift of water into the renal tubes and collecting ducts increases renal NMR intensity in a fashion parallel to overhydration.

Glucagon and Gas

Non-enhanced NMR imaging has shown difficulty in differentiating between bowel loops and adjacent normal and pathological structures. Of particular interest is the need to differentiate between the gastric wall, surrounding fatty tissues, the pancreas and the duodenum. It is known (Stark et al. 1984) that NMR differentiation of the bowel from the pancreas is improved if the bowel lumen is distended with gas or water. By iatrogenically distending the upper GI tract with CO_2 gas (which appears black on NMR images) it is possible to contrast between the duodenum and the pancreatic head and also evaluate gastric wall pathology (Weinreb et al. 1984).

The CO_2 gas is formed in situ by the oral administration of effervescent granules (E-Z Gas), followed by water and intravenous glucagon.

References

Alfidi RJ, Haaga JR, El Yousef SJ (1982) Preliminary experimental results in humans and animals with superconducting whole-body NMR scanner. Radiology 143: 175–181

Baldwin JE, Gill B (1982) Approaches to the preparation of oxygen carriers for use as blood substitutes. Med Lab Sci 39: 45–51

Beall PT (1982) Improved NMR contrast for mouse mammary cancer by safe physiological agents. Physiol Chem Phys Med NMR 14: 399–403

Berliner LJ, Connery BG (1983) Binding subsites in human thrombosis. Biochemistry 22: 369–375

Bernadov J, Martino R, Malet-Martino C et al. (1985) 19F NMR: A technique for metabolism and disposition studies of fluorinated drugs. Trends in pharmacological sciences. In: TIPS current techniques. Elsevier Scientific, Amsterdam, pp 103–105

Brady TJ, Goldman MR, Pykett IL, Hinshaw WS (1982) Proton nuclear magnetic resonance imaging of regionally ischaemic canine hearts: Effects of paramagnetic proton signal enhancement. Radiology 144: 343–347

Brady TJ, Rosen BR, Gold HK et al. (1983) Selective decrease in T_1 relaxation times of infarcted myocardium with the use of manganese labelled monoclonal antibody, antimyosin. Proceedings of the Society of Magnetic Resonance in Medicine. SMRM, Berkeley, CA, p 10

Brasch RC (1983) Work in progress: Methods of contrast enhancement for NMR imaging and potential applications. Radiology 147: 781–788

Brasch RC, Nitecki ED, Brant-Zawadzki MN (1982a) "NSFR" contrast agent for the CNS (abstr). XII Symposium Neuroradiologicum

Brasch RC, Nitecki DE, London D (1982b) Proceedings of the 1st Meeting of the Society of Magnetic Resonance in Medicine. SMRM, Berkeley, CA, pp 25

Brasch RC, London DA, Wesbey GE (1983a) Work in progress: Nuclear magnetic resonance study of paramagnetic nitroxide contrast agent for enhancement of renal structures in experimental animals. Radiology 147: 773–780

Brasch RC, Nitecki DE, Brant-Zawadzki M (1983b) Brain nuclear magnetic resonance imaging enhanced by a paramagnetic nitroxide contrast agent. AJNR 4: 1035–1039

Brasch RC, Nitecki DE, Brant-Zawadzki M et al. (1983c) Brain nuclear magnetic resonance imaging enhanced by a paramagnetic nitroxide contrast agent: Preliminary report. AJR 141: 1019–1023

Brasch RC, Ehman RL, Wesbey GE (1983d) NMR tumour enhancement using a new paramagnetic pyrrolidine contrast agent. Radiology 149 (P): 99

Brasch RC, Weinemann HJ, Wesbey GE (1984) Contrast enhanced NMR imaging: Animal studies using gadolinium DTPA complex. AJR 142: 625–630

Brittenham GM, Farrell DE, Harris JW (1982) Magnetic susceptibility measurements of human iron stores. N Engl J Med 307: 1671–1675

Brown MA et al. (1984) Transition metal-chelate complexes as relaxation modifiers in nuclear magnetic resonance. Med Phys 11: 67–72

Budinger TF (1982) Medical applications of nuclear magnetic resonance scanning: some perspectives in relation to other techniques. In: Witcofski RL, Karstaedt N, Partain CL (eds) NMR imaging: proceedings of an international symposium on NMR imaging. Bowman-Gray School of Medicine at Wake Forest University, Winston-Salem NC, pp 51–64

Buonocore E, Hubner K, Kabalka GH et al. (1985) Potential organ specific contrast agents for liver and spleen: Gadolinium labelled liposomes. Proceedings of the 4th Meeting of the Society of Magnetic Resonance Imaging in Medicine. SMRM, Berkeley, CA, pp 838–839

Burnett KR, Wolf GL, Schumacher HR et al. (1985) Gadolinium oxalate: A prototype agent for contrast enhanced imaging of the liver and spleen with magnetic resonance. Magn Res Imaging 3: 65–71

Bydder GM, Steiner RF, Young IR (1982) Clinical NMR imaging of the brain. AJR 139: 215–236

Carr DH (1984) The use of iron and gadolinium chelates as NMR contrast agents: Animals and human studies. Physiol Chem Phys 16: 137–144

Carr DH, Gadian DG (1985) Contrast agents in magnetic resonance imaging. Clin Radiol 36: 561–568

Carr DH, Brown J, Leung W-L, et al. (1984a) Iron and gadolinium chelates as contrast agents in NMR imaging: Preliminary studies. J Comput Assist Tomogr 8: 385–389

Carr DH, Brown J, Bydder GM et al. (1984b) Intravenous chelated gadolinium as a contrast agent in NMR imaging of cerebral tumours. Lancet I: 484–486

Chan HC, Magin RL, Tompkin WAF et al. (1985) ESR study of the interaction between macrophages and liposomes containing spin labels as NMR contrast agents. Proceedings of the 4th Meeting of the Society of Magnetic Resonance Imaging in Medicine. SMRM, Berkeley, CA, pp 846–847

Chen CW, Cohen JS, Myers CE et al. (1984) Paramagnetic metalloporphyrins as potential contrast agents in NMR imaging. FEBS Lett (1274) 168: 70–74

Chilton HM, Jackels SC, Hinson WH et al. (1984) Use of a paramagnetic substance, colloidal manganese sulphide as an NMR contrast material in rats. J Nucl Med 25: 604–607

Clanton JA, Runge VM, Price AC et al. (1984) Contrast enhanced MRI of brain: Experimental and clinical investigation with gadolinium DTPA. Proceedings of the 3rd Meeting of the Society of Magnetic Resonance Imaging in Medicine. SMRM, Berkeley, CA, pp 157–158

Dreyer JL, Beinert H, Keana JFW et al. (1983) A spin label study of the disposition of the iron-sulphur cluster with respect to the active centre of aconitase. Biochim Biophys Acta 745: 229–236

Engelstad BL, Brasch RC (1984) Pharmaceutical development for MRI. In: James TL, Margulis AR (eds) Biomedical magnetic resonance. Radiology Research Education Foundation, San Francisco, pp 139–156

Faithful NS, Klein J, v d Zee HT et al. (1984) Wholebody oxygenation using intraperitoneal perfusions of fluorocarbons. Br J Anaesth 56: 867–872

Felix R, Schorner W, Laniado M et al. (1985) Brain tumours: MR imaging with gadolinium DTPA. Radiology 156: 681–688

Fiel R, Button T, Mark E et al. (1985) MRI contrast agents: metallo-porphyrins. Proceedings of the 4th Meeting of the Society for Magnetic Resonance in Medicine. SMRM, Berkeley, CA, pp 856–857

Fobben E, Wolf G (1983) Gadolinium-DTPA: a potential NMR contrast agent. Effects upon tissue proton relaxation and cardiovascular function in the rabbit. Invest Radiol 18: 55

Frank JA, House WV, Lauterbur PC et al. (1976) Measurement of proton nuclear magnetic longitudinal relaxation times and water content in infarcted canine myocardium and induced pulmonary injury. Clin Res 24: 217A

Gadian DG, Payne JA, Bryant DJ et al. (1985) Gadolinium-DTPA as a contrast agent in MRI — theoretical projections and practical observations. J Comput Assist Tomogr 9: 242–251

Geraldes CF, Sherry AD, Brown RD et al. (1985) NMR of Mn^{2+} and Gd^{2+} complexes of various polyaxamacrocylic ligands — implications for use as contrast agents in NMR imaging. Proceedings of the 4th Meeting of the Society of Magnetic Resonance in Medicine. SMRM, Berkeley, CA, pp 860–861

Goldman MR, Hinshaw WS, Pohost GM (1982a) Quantification of experimental infarction using NMR imaging and paramagnetic ion contrast enhancement in excised canine heart. Circulation 66: 1012–1016

Goldman MR, Brady TJ, Pykett IL (1982b) Qualification of experimental myocardial infarction using NMR imaging and paramagnetic contrast enhancement in excised canine heart. Radiology 142: 246

Gore JC, Doyle FH, Pennock JM (1983) Relaxation rate enhancement observed in vivo by NMR imaging. In: Partain CL, James AE, Rollo FD (eds) Nuclear magnetic resonance imaging. WB Saunders, Philadelphia pp 94–106

Grief WL, Buxton R, Lauffer RB et al. (1985) Pulse sequence optimisation for paramagnetic contrast imaging using an hepatobiliary agent. Proceedings of the 4th Meeting of the Society of Magnetic Resonance in Medicine. SMRM, Berkeley, CA, pp 864–865

Griffeth LK, Rosen GM, Rauckman EJ (1983) Proceedings of the 2nd Meeting of the Society of Magnetic Resonance in Medicine. SMRM, Berkeley, CA, pp 144–145

Grossman RI, Wolf G, Biery D et al. (1984) Gadolinium enhanced nuclear magnetic resonance images of experimental brain abscess. J Comput Assist Tomogr 8: 204–207

Hambright P, Fawwaz R, Valk P et al. (1975) The distribution of various water soluble radioactive metallo-porphyrins in tumour bearing mice. Bioinorg Chem 5: 87–92

Hansen G, Crooks LE, Davis P (1980) In-vivo imaging of rat anatomy with NMR. Radiology 136: 695–700

Hoffman AK, Henderson AT (1961) A new stable free radical D1-t-butylnitroxide. J Am Chem Soc 83: 4671–4672

Hollis DP, Bulkley BH, Nunnally RL (1979) Effect of manganese ion on the phosphorous nuclear magnetic resonance spectra of the perfused rabbit heart: A possible new membrane probe. Clin Res 26: 240A

Hsia JC, Piette LH (1969) Spin labelling as a general method in studying antibody active sites. Arch Biochem Biophys 129: 296–307

Jackson LS, Nelson JA, Case TA et al. (1985) Manganese protoporphyrin IX: a potential intravenous NMR contrast agent. Invest Radiol 20: 226–229

Jacobsen T, Klaveness J (1985) Magnetic particles as contrast media in MRI. Proceedings of the 4th Meeting of the Society of Magnetic Resonance in Medicine. SMRM, Berkeley, CA, pp 868–869

Josipowicz N, Bonnemain B, Caille JM et al. (1985) Contrast media in MRI. Pharmacokinetic studies of DTPA-Gd and DOTA-Gd in the rabbit. Proceedings of the 4th Meeting of the Society of Magnetic Resonance in Medicine. SMRM, Berkeley, CA, p 870

Keana FW, van Nice FL (1984) Influence of structure on the reduction of nitroxide MRI contrast enhancing agents by ascorbate. Physiol Chem Phys 16: 477–480

Knop RH, Patronas NJ, Cohen JS et al. (1984) Contrast enhancement of experimental tumours in MRI with intravenous paramagnetic metalloporphyrins. Proceedings of the 3rd Meeting of the Society of Magnetic Resonance in Medicine. SMRM, Berkeley, CA, pp 423–424

Knop RH, Naegele M, Patronas N et al. (1985) Gadolinium cryptates as a new class of water soluble MRI contrast agents: Comparison to Gd-DTPA for organ pathology localization. Proceedings of the 4th Meeting of the Society of Magnetic Resonance in Medicine. SMRM, Berkeley, CA, pp 871–872

Kosman DJ, Hsia JC, Piette LH (1969) ESR probing of macromolecules: Function and operation of structural units within the active site of crymotrypsin. Arch Biochem Biophys 133: 29–37

Koutcher JA, Burt CT, Lauffer RB et al. (1984) Contrast agents and spectroscopic probes in NMR. J Nucl Med 25: 506–513

Lanaido M, Weinmann HJ, Schorner W et al. (1984) First use of Gd-DTPA/dimeglumine in man. Physiol Chem Phys 16: 157–165

Lauffer RB (1984) The design of NMR contrast agents. Proceedings of the 3rd Meeting of the Society of Magnetic Resonance in Medicine. SMRM, Berkeley, CA, pp 446–448

Lauffer RB, Greif WL, Stark DD et al. (1985a) Iron EHPG as an hepatobiliary MR contrast agent. Initial imaging and biodistribution studies. J Comput Assist Tomogr 9: 431–438

Lauffer RB, Greif WL, Stark DD et al. (1985b) Iron EHPG as an hepatobiliary MR contrast agent. Initial imaging and biodistribution studies. Proceedings of the Meeting of the Society of Magnetic Resonance in Medicine. SMRM, Berkeley, CA, pp 883–884

Lauterbur PC, Mendonca-Dias MH, Rudin AM (1978a) Augmentation of tissue water proton spin-lattice relaxation rates by in-vivo addition of paramagnetic ions. In: Dutton PL, Leigh JC, Scarpa A (eds) Frontiers of biological energetics, vol. 1. Academic Press, New York, pp 752–759

Lauterbur PC, Jacobson MJ, Rudin AM (1978b) Augmentation of the water proton spin-lattice relaxation in tissue by in-vivo injection of manganous ion. Abstracts of the 19th Annual Meeting of the European Nuclear Conference. B19

Letokhov VS (1985) Laser biology and medicine. Nature 13 (July 25): 325–330

London DA, Davis PL, Williams RD et al. (1983) Nuclear magnetic resonance imaging of induced renal lesions. Radiology 148: 167–172

Lyon R, Faustino P, Mornex F et al. (1985) Tissue distributions and stability of metalloporphyrin NMRI contrast agents. Proceedings of the 4th Meeting of the Society of Magnetic Resonance in Medicine. SMRM, Berkeley, CA, pp 885–886

Mattrey RF, Long DM, Multer F et al. (1982) Perfluoroctylbromide: A reticuloendothelial-specific and tumour-imaging agent for computed tomography. Radiology 145: 755–758

McNamara MT, Higgins CB, Ehman RL et al. (1984) Acute myocardial ischemia: magnetic resonance contrast enhancement with gadolinium-DTPA. Radiology 153: 157–163

McNamara MT, Wesbey GE, Brasch RC et al. (1985) MRI of acute myocardial infarction using a nitroxyl spin label PCA. Invest Radiol 20: 591–595

Mela L, Chance B (1965) Spin labelled biomolecules. Proc Natl Acad Sci USA 54: 1010

Menasche P, Fauchet M, Lavergne A et al. (1984) Reduction of myocardial infarct size by a fluorocarbon-oxygenated perfusate. Am J Cardiol 53: 608–613

Mendonca-Dias MH, Gaggelli E, Lauterbur PC (1983) Paramagnetic contrast agents in nuclear magnetic resonance medical imaging. Semin Nucl Med 13: 364–376

Mendonca-Dias MH, Bernardo ML, Muller RN et al. (1985) Ferromagnetic particles as contrast agents for magnetic resonance imaging. Proceedings of the 4th Meeting of the Society of Magnetic Resonance in Medicine. SMRM, Berkeley, CA, pp 887–888

Newhouse JH, Pykett IL, Brady TJ (1981) NMR scanning of the abdomen: preliminary results in small animals. In: Witcofski RL, Karstaedt N, Partain CL (eds) NMR imaging: proceedings of an international symposium on NMR imaging. Bowman-Gray School of Medicine Press, Winston-Salem, NC, pp 121–124

Newhouse JH, Brady TJ, Gebhardt M (1982) NMR imaging: preliminary results in the upper extremities of man and the abdomen of small animals. Radiology 142: 246

Ohgushi M, Nagayoma K, Wada A (1978) Destran magnetitite: a new relaxation reagent and its applications to T_2 measurements in gel systems. J Magn Res 29: 599–601

Olsson M, Persson BRB, Salford LG et al. (1985) Ferromagnetic particles as contrast agent in T_2 NMR-imaging. Proceedings of the 4th Meeting of the Society of Magnetic Resonance in Medicine. SMRM, Berkeley, CA, p 889

Papahadjopopulos D (ed) (1978) Liposomes and their uses in biology and medicine. Ann NY Acad Sci 308: 1–462

Patronas NJ, Hekmatpanah J, Doi K (1983) Brain tumour imaging using radiopaque perfluorocarbons — a preliminary report. J Neurosurg 58: 650–653

Patronas NJ, Knop RH, Hanbright R et al. (1985) Paramagnetic labelled hematoporphyrin derivative (HPD) as a human tumour localizing agent for laser phototherapy directed treatment for MRI images. Proceedings of the 4th Meeting of the Society of Magnetic Resonance in Medicine. SMRM, Berkeley, CA, pp 890–891

Perkins RC, Beth AH, Wikerson LS (1980) Enhancement of free radical reduction by elevated concentrations of ascorbic acid in avian dystrophic muscle. Proc Natl Acad Sci USA 77: 790–794

Pohost GM, Goldman MR, Pykett IL (1982) Gated NMR imaging in canine myocardial infarctions. Circulation, Part II, 66: Communications 155: 1015

Pople JA, Schneider WG, Bernstein HJ (1959) High resolution NMR. McGraw Hill, New York, p 209

Pykett IL, Rosen BR (1983) Nuclear magnetic resonance: in-vivo proton chemical shift imaging: Work in progress. Radiology 149: 197–201

Runge VM, Stewart RG, Clanton JA et al. (1983a) Work in progress: Potential oral and intravenous paramagnetic NMR contrast agents. Radiology 147: 789–791

Runge VM, Clanton JA, Smith FW et al. (1983b) Nuclear magnetic resonance of iron and copper disease states. AJR 141: 943–948

Runge VM, Clanton JA, Lukehart CM, (1983c) Paramagnetic agents for contrast enhanced NMR imaging: A review. AJR 141: 1209–1215

Runge VM, Clanton JA, Price AC et al. (1984a) Paramagnetic contrast agents in magnetic resonance imaging: Research at Vanderbilt University. Physiol Chem Phys Med NMR 16: 113–122

Runge VM, Clanton JA, Herzer WA et al. (1984b) Intravascular contrast agents suitable for magnetic resonance imaging. Radiology 153: 171–176

Runge VM, Foster MA, Clanton JA et al. (1984c) Contrast enhancement of magnetic resonance images by chromium EDTA: An experimental study. Radiology 152: 123–126

Runge VM, Price AC, Wehr C (1985a) Contrast enhanced MRI: evaluation of a canine model of osmotic blood–brain barrier disruption. Invest Radiol 20: 830—844

Runge VM, Foster MA, Clanton JA et al. (1985b) Particulate oral NMR contrast agents. Int J Nucl Med Biol 12: 37–42

Saini S, Widder D, Stark DD et al. (1985) MR contrast enhancement in detection of liver tumours using tissue specific reticuloendothelial agents. Proceedings of the 4th Meeting of the Society of Magnetic Resonance in Medicine. SMRM, Berkeley, CA, pp 896–897

Schorner W, Felix R, Laniado M et al. (1984a) Gadolinium-DTPA an Menschen. Verträglichkeit, Kontrastbeeinflussung und erste klinische Ergebnisse. ROFO 140: 493–500

Schorner W, Laniado M, Felix R (1984b) Erster klinischer Einsatz von Gadolinium-DTPA in der kernspintomographischen Darstellung einer paradelvinen Nierenzyste. ROFO 141(2): 227–228

Schumacher JH, Matys ER, Clorius JH et al. (1985) Contribution of paramagnetic trace elements to the spin-lattice relaxations time in the live. Invest Radiol 20: 601–608

Sharma P, Tyrell DA, Ryman BE (1977) Some properties of liposomes of different sizes. Biochem Soc Trans 5: 1146–1149

Sohn M, Roberts MF, Spokojny AM et al. (1985) Metalloporphyrins: MRI contrast agents for atherosclerotic vascular disease. Proceedings of the 4th Meeting of the Society of Magnetic Resonance in Medicine. SMRM, Berkeley, CA, pp 902–903

Spector WS (ed) (1956) Handbook of toxicology. WB Saunders, Philadelphia

Stacy BD, Thorburn GD (1966) 51-Chromium-EDTA for estimation of glomerular filtration rate. Science 152: 1076–1077

Stark DD, Moss AA, Goldberg HI et al. (1984) Magnetic resonance and CT of the normal and diseased pancreas: a comparative study. Radiology 150: 153–162

Stone TJ, Buckman T, Nordio PL (1965) Spin labelled biomolecules. Proc Natl Acad Sci USA 54: 1010–1017

Strich G, Hagan PL, Gerber KH, Slutsky RA (1985) Tissue distribution and magnetic resonance spin lattice relaxation effects of gadolinium DTPA. Radiology 154: 723–726

Swartz HM, Chen K, Pals M et al. (1986) Hypoxia sensitive NMR contrast agents. Magn Reson Med 3: 169–174

Unger EC, Totty WG, Neufeld DM et al. (1985) Magnetic resonance imaging using gadolinium labelled monoclonal antibody. Invest Radiol 20: 693–700

Weinmann HJ, Brasch RC, Press WR et al. (1984a) Characteristics of gadolinium-DTPA complex: a potential NMR contrast agent. AJR 142: 619–624

Weinmann HJ, Laniado M, Muetzel W (1984b) Pharmacokinetics of Gd-DTPA/dimeglumine after intravenous injection into healthy volunteers. Physiol Chem Phys Med NMR 16: 167–172

Weinreb JC, Maravilla KR, Redman HC et al. (1984) Improved MR imaging of the upper abdomen with glugagon and gas. J Comput Assist Tomogr 8: 835–838

Wesbey GE, Brasch RC, Engelstad B et al. (1983a) Gastrointestinal contrast enhancement for NMR imaging using non-toxic oral iron solutions. Invest Radiol 18: S4

Wesbey GE, Brasch RC, Engelstad BL (1983b) Nuclear magnetic resonance contrast enhancement study of the gastrointestinal tract of rats and a human volunteer using non-toxic oral iron solutions. Radiology 149: 175–180

Wesbey GE, Engelstad BL, Brasch RC (1984a) Paramagnetic pharmaceuticals for magnetic resonance imaging. Physiol Chem Phys 16: 145–155

Wesbey GE, Higgins CB, McNamara MT et al. (1984b) Effect of gadolinium DTPA on the magnetic relaxation times of normal and infarcted myocardium. Radiology 153: 165–169

White DL, Ramos EC, Huberty JP et al. (1985) Iron (II) complexes as MRI contrast agents. Proceedings of the 4th Meeting of the Society of Magnetic Resonance in Medicine. SMRM, Berkeley, CA, pp 906–907

Wolf GL, Baum L (1983) Cardiovascular toxicity and tissue proton T_1 response to manganese injections in the dog and rabbit. AJR 141: 193–197

Wolf GL, Fobben ES (1984) Tissue proton T_1 T_2 response to gadolinium DTPA injection to rabbits: a potential renal contrast agent for NMR imaging. Invest Radiol 19: 324–328

Wolf GL, Burnett KR, Goldstein EJ et al. (1985) Contrast agents for magnetic resonance imaging. Magnetic resonance annual. Raven Press, New York

Young IR, Clarke GJ, Bailes DR (1981) Enhancement of relaxation rate with paramagnetic contrast agents in NMR imaging. Computed Tomography 5: 543–554

Young IR, Bailes DR, Collins AG et al. (1982) Initial clinical evaluation of a wholebody NMR tomograph. J Comput Assist Tomogr 6: 1–18

Zanelli GD, Kaelin AC (1981) Synthetic porphyrins as tumour localising agents. Br J Radiol 54: 403–407

8. Clinical Use of Intravenous Gadolinium-DTPA in Magnetic Resonance Imaging of the Central Nervous System

G.M. Bydder, H.P. Niendorf and I.R. Young

Introduction

Following extensive pharmacological and animal studies (Weinmann and Grier 1983; Weinmann et al. 1984), clinical trials with intravenous gadolinium-DTPA (Gd-DTPA) were begun at the end of 1983 (Shorner et al. 1984). During 1984 one group in Berlin (Claussen et al. 1985; Felix et al. 1984a, b) and another in London (Carr et al. 1984a, b; Graif et al., to be published; Curati et al., to be published; Bydder et al. 1985) conducted clinical trials. During 1985 Gd-DTPA was released more widely and now over 20 groups in Europe and four in the United States are involved in clinical evaluation programmes. Gd-DTPA is the first parenteral magnetic contrast agent available for clinical use and the results of these clinical trials have been followed with considerable interest.

In general terms the clinical results with Gd-DTPA parallel those seen with iodinated contrast agents used in X-ray computed tomography (CT) but there are important differences, particularly in relation to the use of pulse sequences and effects due to increased concentration of Gd-DTPA.

In this chapter the mode of action of Gd-DTPA is reviewed and the clinical results in the central nervous system are summarised.

Paramagnetic Contrast Agents

The MRI features of paramagnetic agents have been extensively reviewed, including systematic analysis of their mechanism of action (Koenig et al., to be published; Koenig and Brown, to be published; Gadian et al. 1985; Brasch 1983; Runge et al. 1983).

There are a variety of naturally occurring contrast agents. These include molecular O_2 (Tripathi et al. 1984), iron in the form of haemosiderin and ferritin, Fe^{3+} and methaemoglobin. The effect of O_2 is small but effects attributable to Fe^{3+} and iron-containing compounds are not uncommon. These are important in cases of systemic iron overload (Leung et al. 1984) and Hallervorden-Spatz disease as well as in explaining the appearance of haemorrhage.

The basic pharmacology of Gd-DTPA has been studied by Wienmann et al. (1983, 1984). Gd-DTPA has a molecular weight of 690, which is similar to that of iodinated contrast agents. After intravenous injection it circulates within the vascular system but does not cross the normal blood–brain barrier. It is excreted unchanged through the kidney and accumulates in the urine. The LD_{50} compares favourably with both iodinated contrast agents and other chelated complexes of Gd such as Gd-ETPA. The half-time of Gd-DTPA is relatively short, being of the order of 20 min in animal studies, but rather longer in human studies.

Gd-DTPA crosses the abnormal blood–brain barrier. Its accumulation therefore tends to parallel that of the iodinated contrast agents used in CT. Areas where there is no blood–brain barrier such as the falx and the choroid plexus show enhancement but the level of enhancement within the brain is small and attributable to accumulation of the Gd-DTPA in blood vessels. Thus grey matter shows greater enhancement than white matter, in keeping with its greater vascularity, although the degree of enhancement of both tissues is quite small.

Effects of Paramagnetic Contrast Agents on NMR Parameters

Effects of Gd-DTPA on Tissues T_1 and T_2

The effect of paramagnetic agents on solutions has been well studied since the earliest days of NMR and the mechanism of action of paramagnetic ions is reasonably well understood.

In general, relaxation rates (i.e. the reciprocal of relaxation times) are additive so that the observed T_1 T_{10} of a solution is given by

$$\frac{1}{T_{10}} = \frac{1}{T_{1I}} + \frac{1}{T_{1P}} \tag{1}$$

where T_{1I} is the initial spin-lattice relaxation of the solution and T_{1P} is the spin-lattice relaxation time of the paramagnetic agent. The same relationship is true for T_2.

Now the relaxation rate of a paramagnetic agent in solution is proportional to the concentration of that agent so that for Gd-DTPA

$$\frac{1}{T_{10}} = \frac{1}{T_{10}} + k\,[\text{Gd-DTPA}] \tag{2}$$

where T_{10} is the relaxation time in the absence of Gd-DTPA and k is a constant. Two important consequences follow, the first of which is illustrated in Fig. 8.1. For the same concentration of Gd-DTPA the reduction in T_1 or T_2 is greater for tissues with longer values of T_1 or T_2 than it is for shorter values, i.e. the relaxations of tissues

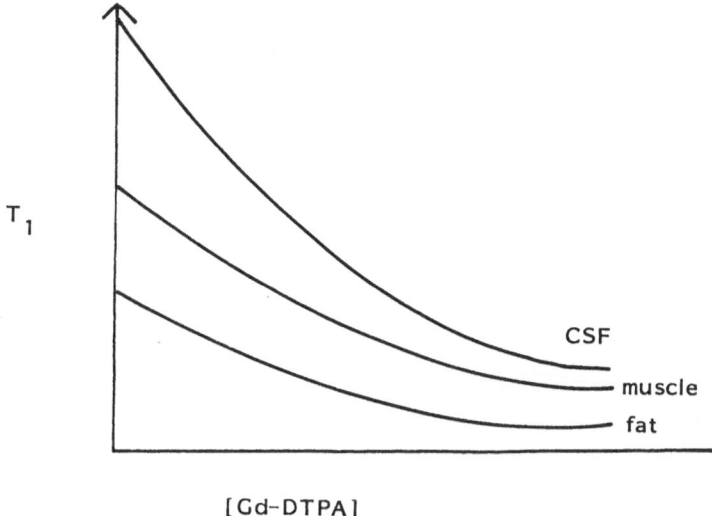

T_1

CSF

muscle

fat

[Gd-DTPA]

Fig. 8.1.   Change in T_1 with increasing concentration of Gd-DTPA (schematic) for CSF, muscle and fat.

tend to converge with increasing concentration. The second consequence follows because T_2 is always smaller than T_1 for biological tissues and often several times smaller, and hence the absolute effect on T_1 is greater than that on T_2.

Effect of Different Pulse Sequences

While the changes in T_1 and T_2 produced by Gd-DTPA discussed in the previous section are relatively straightforward, they are complicated in practice by the fact that the pulse sequences used in everyday practice are dependent on both T_1 and T_2, and that in general a decrease in T_1 tends to *increase* relative signal intensity whilst a decrease in T_2 tends to *decrease* relative signal intensity. The net effect therefore depends in detail on the degree of T_1 and T_2 dependence of the pulse sequence for the particular tissue under consideration.

At low concentrations of Gd-DTPA the increase in signal produced by Gd-DTPA is usually the dominant effect while at high concentrations the decrease in signal intensity produced by the effect on T_2 is usually dominant. Thus the change in signal intensity produced by Gd-DTPA goes through a maximum before decreasing. This maximum is highest for highly T_1 dependent sequences such as inversion recovery and least for highly T_2 dependent sequences (Fig. 8.2).

Net Effect of Administration of Gd-DTPA

The observed effects of Gd-DTPA depend on the dosage. In this study 0.1 mmol/kg was used but in other studies dosages of 0.2 mmol/kg have been used. The net effect also depends on the differences between normal and abnormal tissues before and after examinations as well as the pulse sequence. In addition to the partial saturation

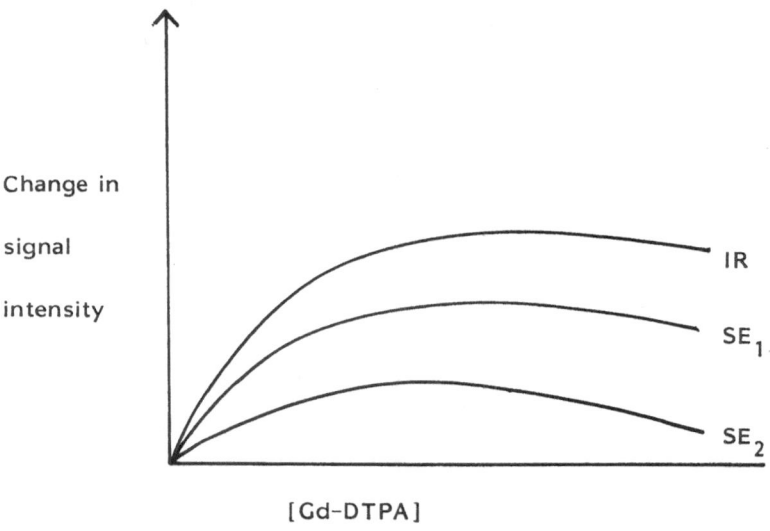

Fig. 8.2. Change in signal intensity of a tissue with increasing concentration of Gd-DTPA with inversion recovery (*IR*), short TE short TR (SE$_1$) and long TE long TR (SE$_2$) sequences (schematic).

(PS), inversion recovery (IR) and spin echo (SE) sequences it is possible to compute T$_1$ and T$_2$ maps which show these parameters alone.

Although the overall effects are complex and depend on dosage, time of administration, pulse sequence and tissue type, the behaviour of Gd-DTPA has been consistent enough for a number of generalisations to be made for several different clinical conditions. Some of these are discussed below.

Results of Paramagnetic Contrast Agents in Clinical Studies

Acoustic Neuroma

There have been several studies of acoustic neuromas with unenhanced MRI which have demonstrated tumours with a high level of accuracy so that the role of Gd-DTPA in MRI is likely to be less than that of iodinated contrast agents used with CT, where contrast enhancement is used in virtually all cases of acoustic neuroma and frequently adds significant new information.

In all 17 cases so far examined with Gd-DTPA enhanced MRI, significant contrast enhancement has been seen. This has been greatest with the inversion recovery sequence (IR$_{1500/500/44}$), intermediate with the short TE short TR spin echo sequence (SE$_{544/44}$), and least with the long TE long TR sequence (SE$_{1500/80}$). In terms of efficacy, contrast enhancement has been of significant value in two cases of small acoustic neuroma where the diagnosis was not certain in the pre-enhancement images (Fig. 8.3). It was also of value in one case of tumour recurrence where the anatomy was distorted by the previous surgery. These results are presented in more detail elsewhere (Curati et al. 1985).

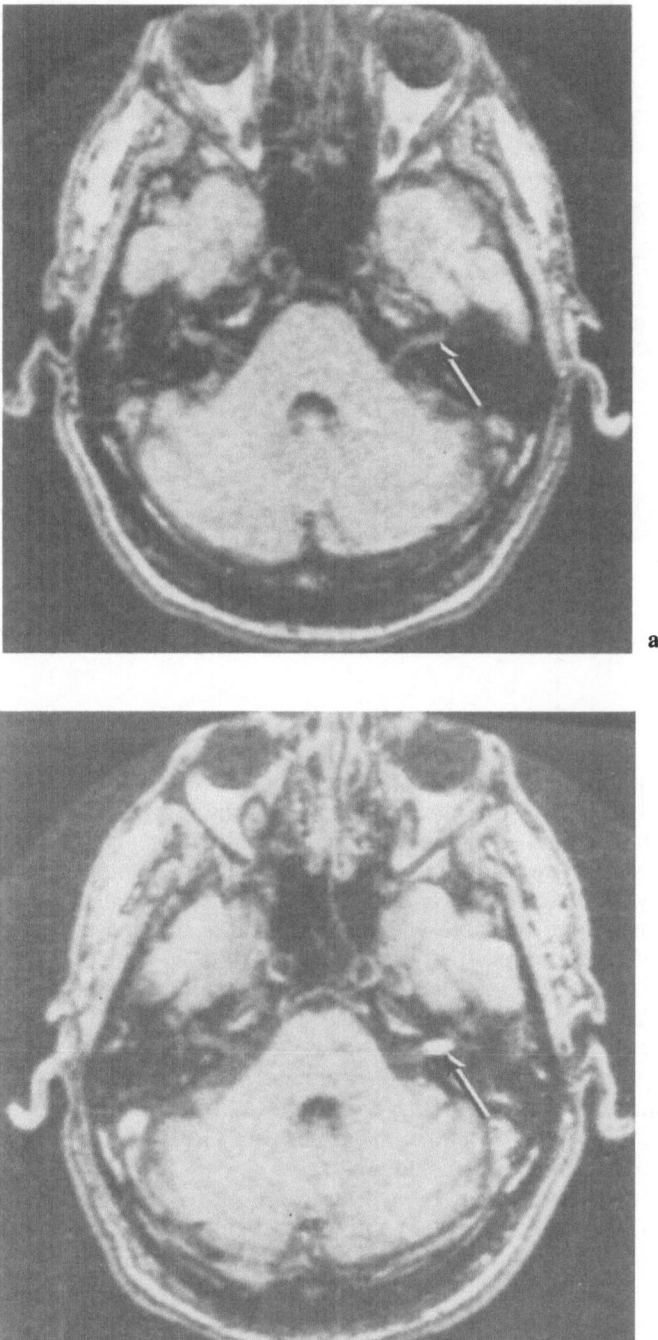

Fig. 8.3. Acoustic neuroma: PS_{300} sequences **a** before Gd-DTPA and **b** after Gd-DTPA. The small intracanalicular tumour is highlighted in **b**.

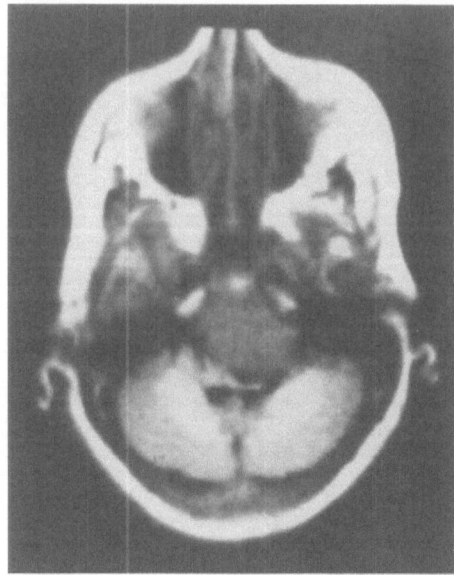 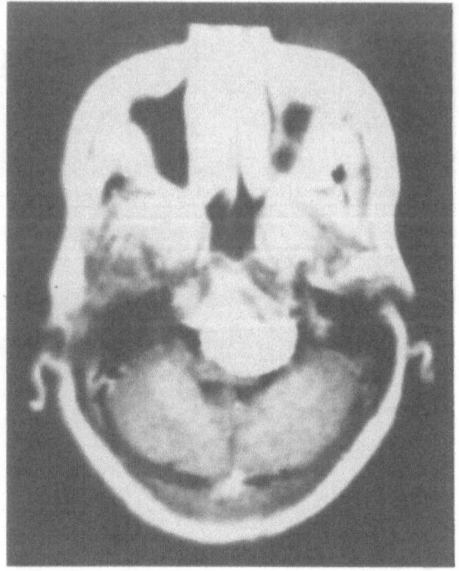

a   b

Fig. 8.4. Clivus meningioma: IR$_{1500/500/44}$ scans **a** before Gd-DTPA and **b** after Gd-DTPA. A high level of enhancement is seen.

Meningioma

There have been two published reports with unenhanced MRI in which this technique was found to be inferior to contrast enhanced X-ray CT (Bradley et al. 1984; Zimmerman 1984). The high level of contrast enhancement usually seen with X-ray CT is a valuable sign in recognising meningiomas and the absence of the equivalent feature in early MRI studies was a distinct disadvantage, particularly as the degree of contrast enhancement available with meningioma is frequently greater than that seen with malignant tumours.

Our initial experience with Gd-DTPA has shown contrast enhancement in each case (e.g. Fig. 8.4) and the results overall have compared well with X-ray CT (Bydder et al. 1985).

Magnetic resonance imaging has been of particular value in defining the relationship of the tumour to the tentorium (Fig. 8.5) as well as distinguishing tumour from oedema (Fig. 8.6). This latter case also illustrates that Gd-DTPA can be of particular value in cases of meningioma en plaque.

Further studies with unenhanced MRI have shown improved results in the detection of meningioma (Bilaniuk et al. 1985) and contrast enhanced MRI has also compared well with X-ray CT in another study.

Other Benign Tumours

A small number of pituitary tumours have been studied and all of these have displayed contrast enhancement (e.g. Fig. 8.7). Other tumours including glomus

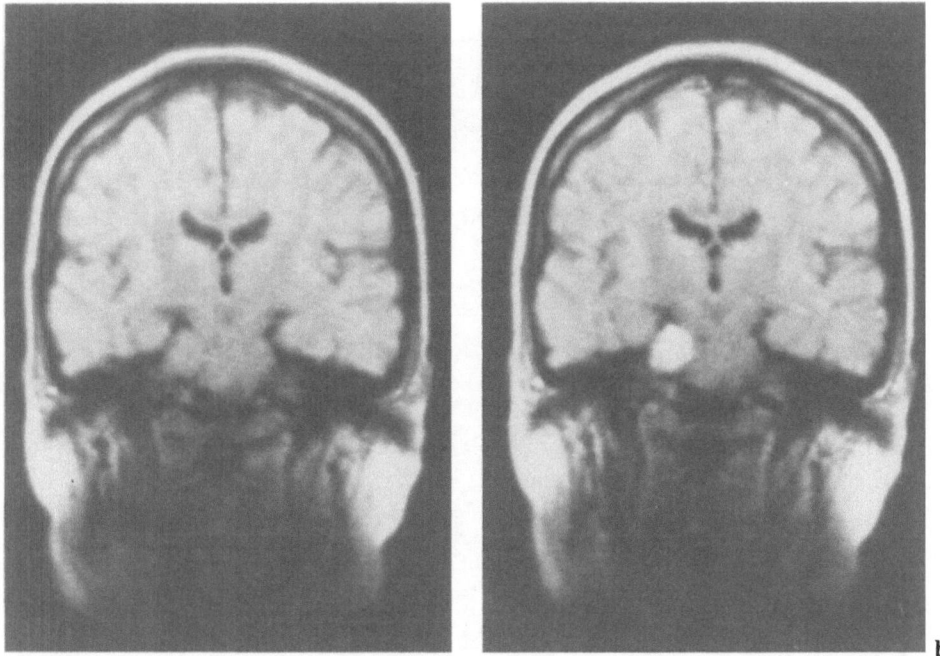

Fig. 8.5. Subtentorial meningioma: $SE_{544/44}$ scans **a** before Gd-DTPA and **b** after Gd-DTPA. A moderate level of enhancement is seen.

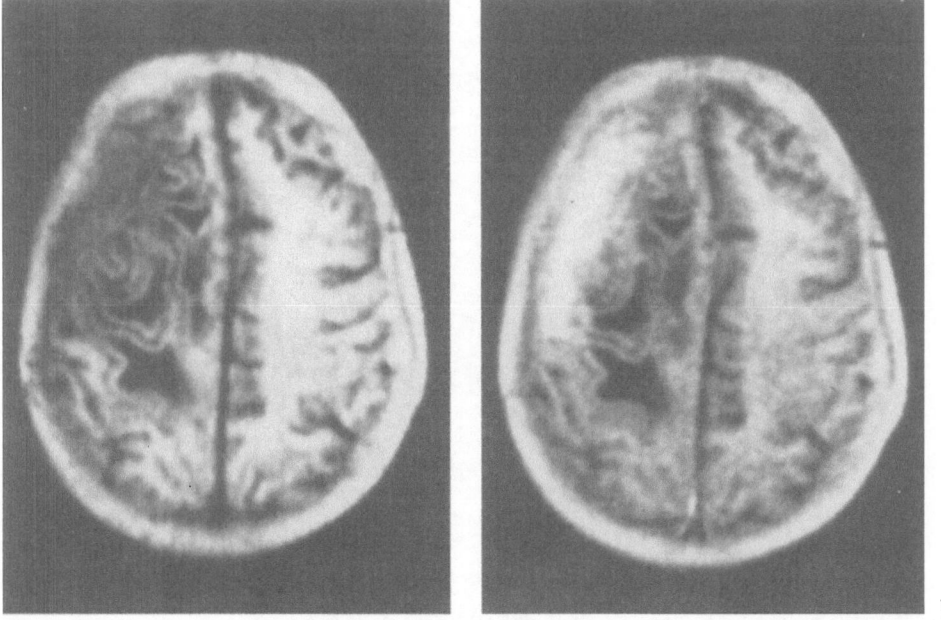

Fig. 8.6. Meningioma en plaque: $IR_{1500/500/44}$ scans **a** before Gd-DTPA and **b** after Gd-DTPA. The tumour and oedema are distinguished in **b**.

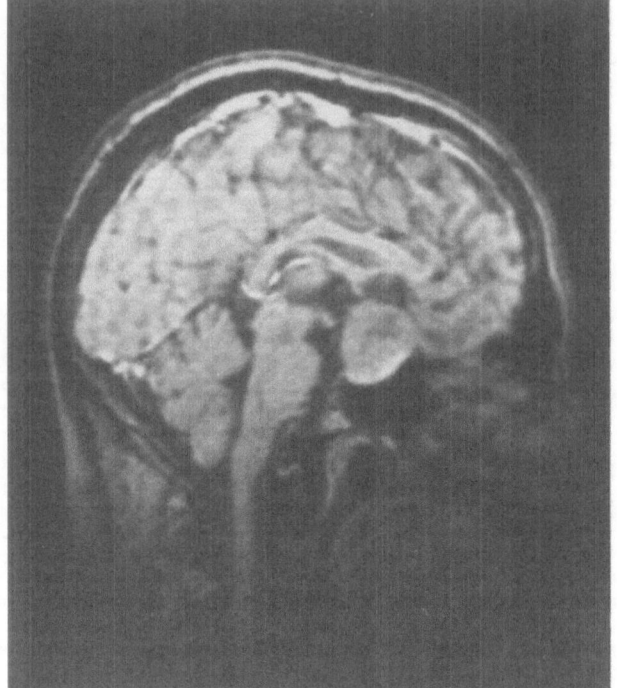

a

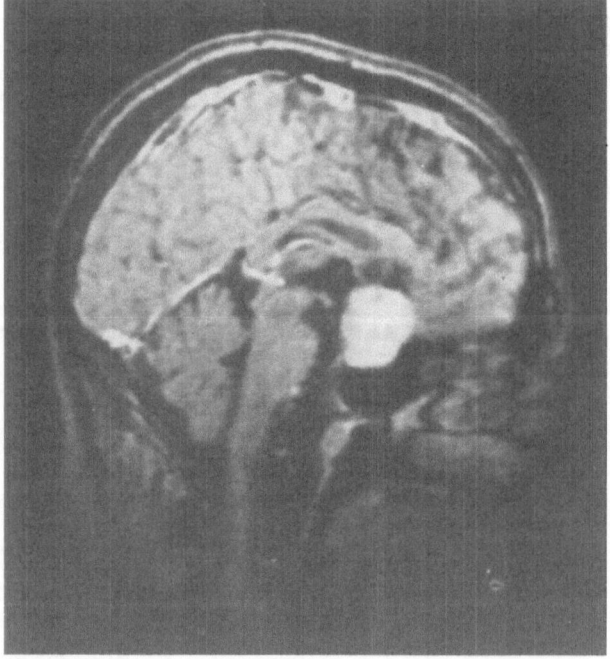

b

Fig. 8.7. Pituitary adenoma: $PS_{500/22}$ scans **a** before Gd-DTPA and **b** after Gd-DTPA. Moderate enhancement of the tumour is seen in **b**.

jugulare tumours, a chordoma, a schwannoma and an epidermoid tumour have also displayed enhancement. In these cases the enhancement has been useful in distinguishing cystic from solid lesions as well as defining the boundaries of the tumours more precisely. More work is required before any definitive assessment can be made.

Malignant Tumours

This has been the group of diseases studied in most detail to date with studies published by both the Berlin (Claussen et al. 1985) and London (Graif et al., to be published) groups and major trials are underway in the United States.

Again the results tend to parallel those seen with iodinated agents in CT. Tumours of higher grades of malignancy generally show more contrast enhancement than tumours of lower grade. Areas of cystic necrosis may also show enhancement but the greatest clinical value of Gd-DTPA has been in distinguishing tumour from oedema (Fig. 8.8). This has been previously perceived as one of the major disadvantages of MRI in comparison with CT in early clinical studies.

As in other studies there is a profound difference related to the sequence used. For example the $IR_{1500/500/44}$ images display multiple ring enhancement in Fig. 8.9. These changes are difficult to perceive on the $SE_{1500/80}$ sequences performed at the same level in the same figure.

Tumours which do not display an increase in T_1 and T_2 have attracted considerable attention recently (McKay et al. 1985); there are potential difficulties in their

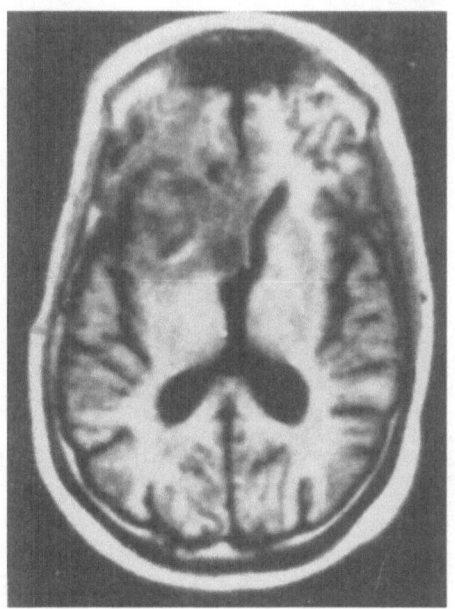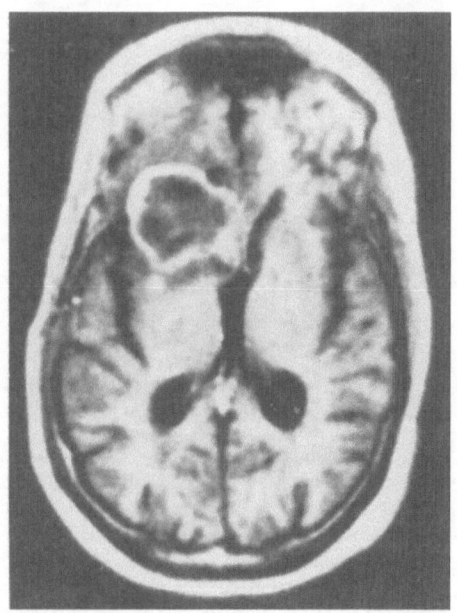

a   b

Fig. 8.8. Astrocytoma grade IV: $IR_{1500/500/44}$ scans **a** before Gd-DTPA and **b** after Gd-DTPA. Ring enhancement is seen.

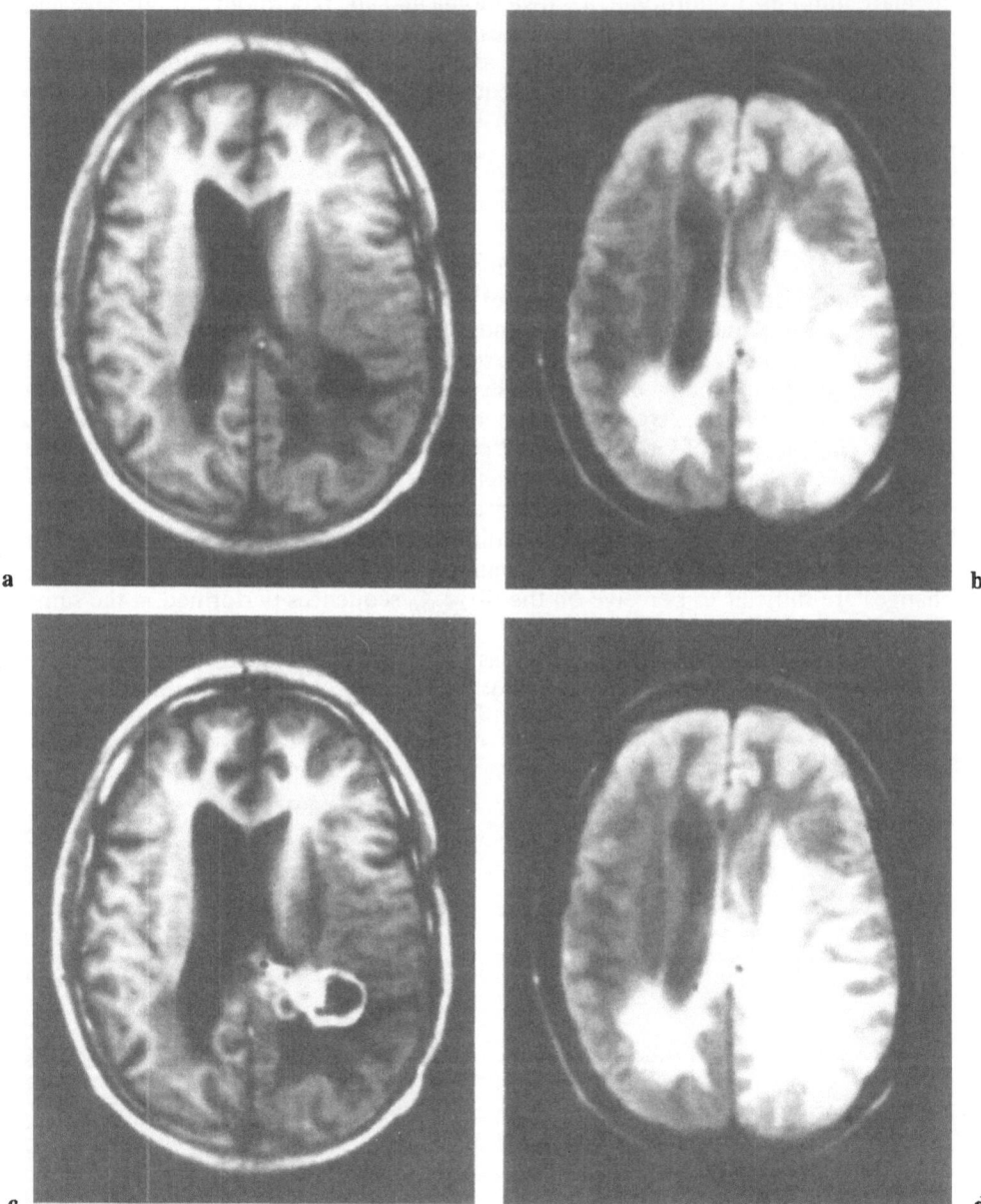

Fig. 8.9. Ependymoma: **a** $IR_{1500/500/44}$ and **b** $SE_{1500/80}$ scans before Gd-DTPA and **c** $IR_{1500/500/44}$ and **d** $SE_{1500/80}$ scans after Gd-DTPA. Multiple ring enhancement is seen in **c** but poorly shown in **d**.

recognition and indirect signs are frequently more important than differences in contrast in their diagnosis. Figure 8.10 is an example of such a tumour. It is a metastasis from a carcinoma of the rectum which is poorly shown with the $IR_{1500/500/44}$ sequence although it is reasonably well defined after intravenous Gd-DTPA. The tumour is

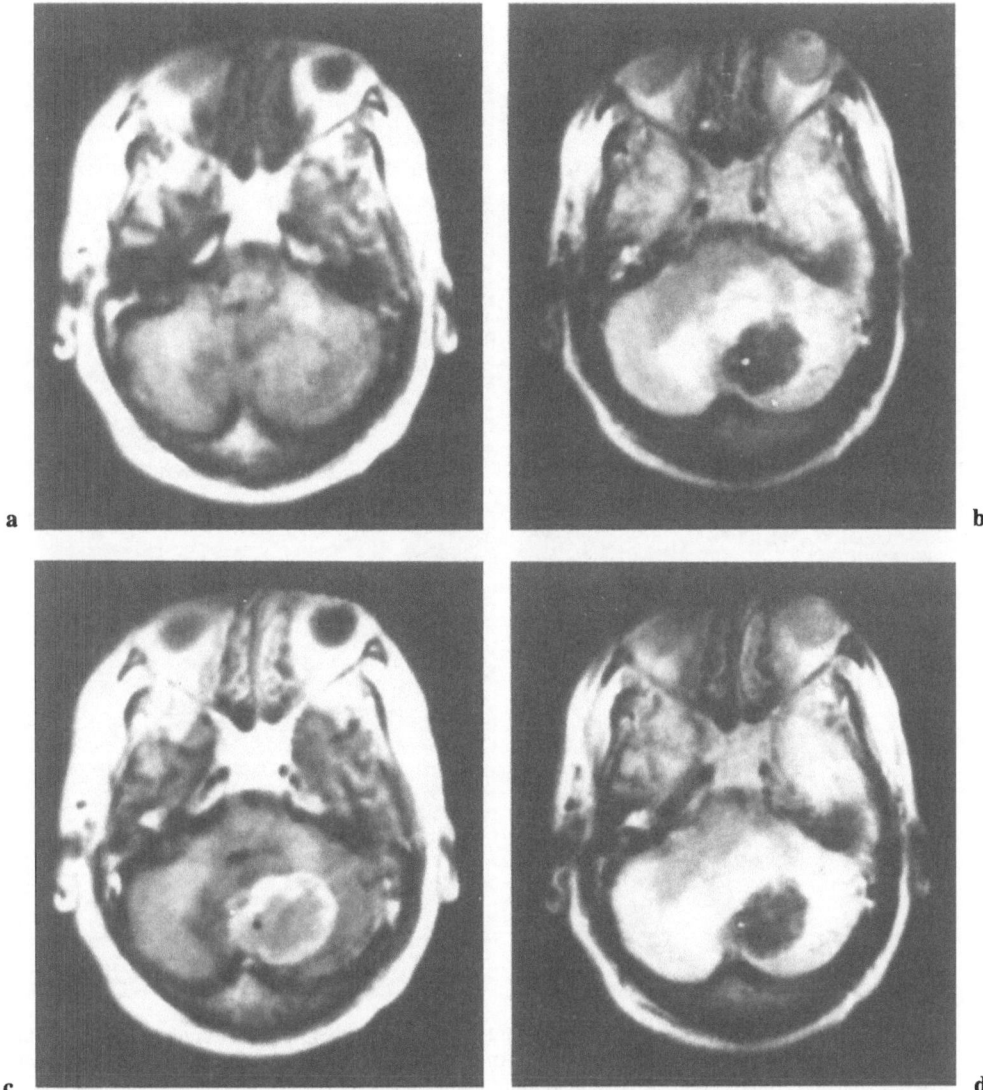

a   b

c   d

Fig. 8.10.   Metastasis from carcinoma of the rectum: **a** IR$_{1500/500/44}$ and **b** SE$_{1500/80}$ scans before Gd-DTPA compared with **c** IR$_{1500/500/44}$ and **d** SE$_{1500/80}$ scans after Gd-DTPA. Ring enhancement is seen in **c** but not in **d**.

well demonstrated with the SE$_{1500/80}$ sequence because its T$_2$ is shorter than that of brain.

Comparison with CT has generally been favourable in terms of the degree of enhancement, distinction between tumour and oedema, and extent of abnormality. Some areas of apparent oedema on the pre-enhancement scan may show contrast enhancement indicating that they are probably areas of tumour infiltration. In these cases this infiltration has not been apparent with CT (e.g. Fig. 8.11).

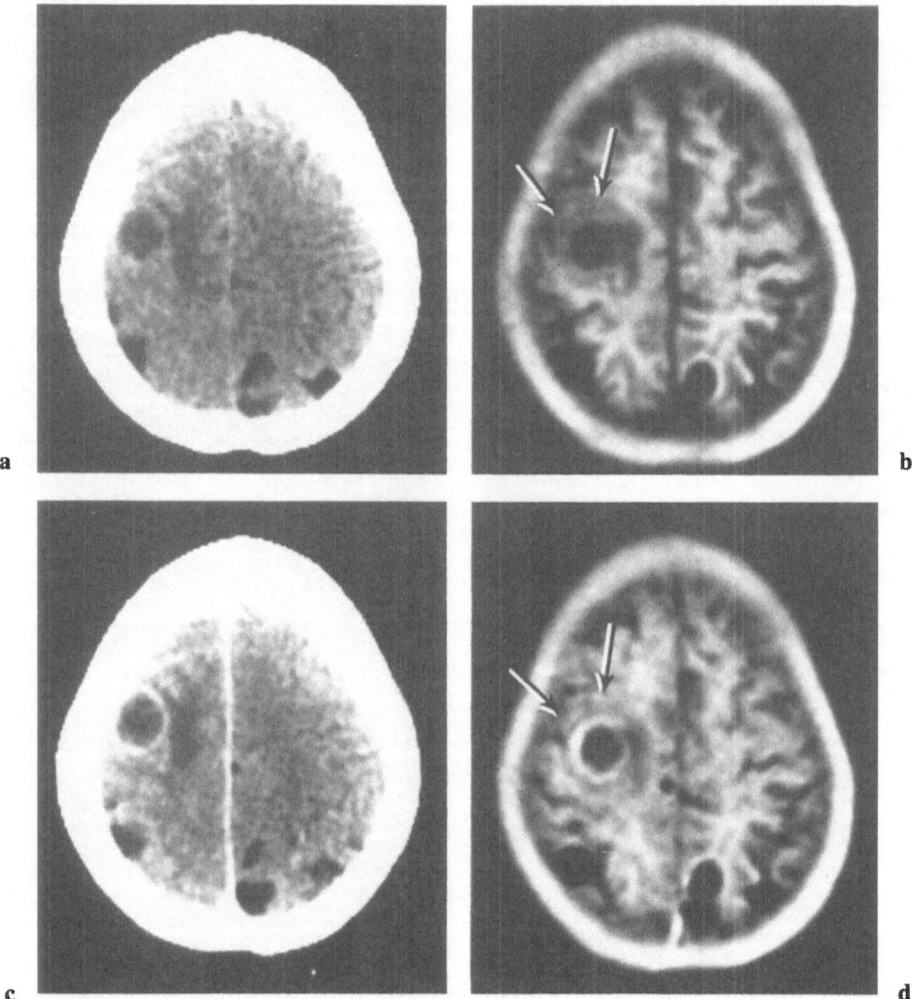

Fig. 8.11. Metastasis from breast Ca: **a** CT and **b** IR$_{1500/500/44}$ scans before enhancement compared with **c** CT and **d** IR$_{1500/500/44}$ scans after enhancement. An area of apparent "oedema" in **b** shows enhancement in **d** (*arrows*) and probably represents tumour infiltration. No change is seen in this area in **c**.

Use in Non-tumourous Disease

Experience in neurological non-tumourous disease is still small but several studies are now underway and some tentative comment can be made.

Enhancement has been seen in some lesions of multiple sclerosis (e.g. Fig. 8.12). There is general agreement that MRI is more sensitive than CT in this disease. The sensitivity of CT has been improved by use of double dose contrast enhancement. There may also be some improvement in sensitivity with MRI using contrast enhancement. It is also possible that contrast enhancement will be of value in distinguishing between "active" and "inactive" lesions. Contrast enhancement has been seen around infarction, in arteriovenous malformations and in the wall of

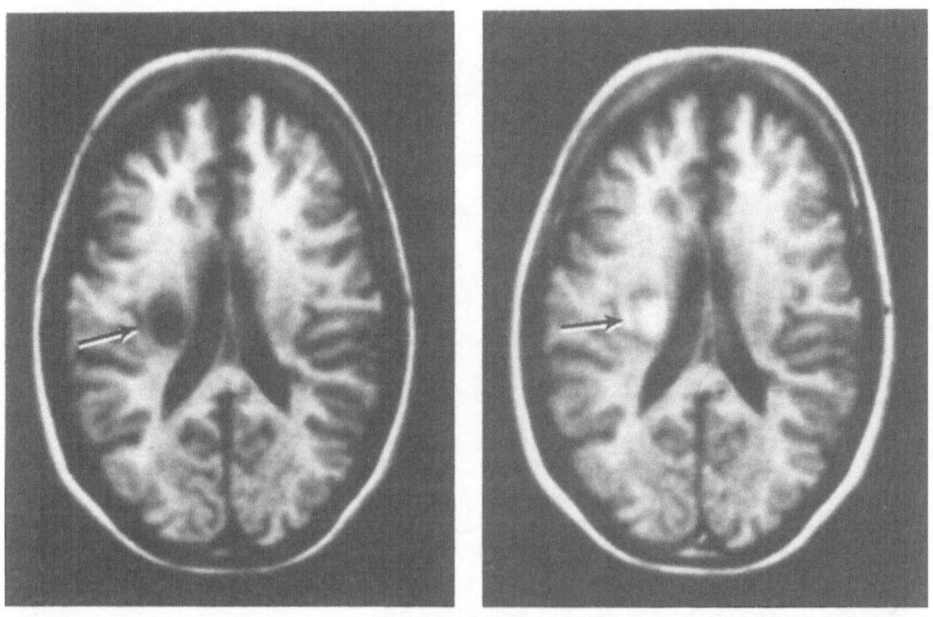

a   b

Fig. 8.12. Multiple sclerosis: $IR_{1500/500/44}$ scans **a** before Gd-DTPA and **b** after Gd-DTPA. A lesion displays enhancement (*arrows*).

aneurysms but, as with CT, the contribution of contrast enhancement may be relatively limited.

In two cases of infection contrast enhancement has been seen. As with tumours, distinction between the central lesion and the surrounding oedema has been the most important single feature in these cases.

Orbit, Nasopharynx and Neck

Enhancing lesions have been demonstrated in the orbit in two instances and it is possible that useful diagnostic information may be obtained in a range of diseases in the orbit.

The nasopharynx is of particular interest. The normal mucosa shows enhancement and some tumour enhancement has been seen in a limited range of cases. This has been of value in demonstrating invasion through the cribiform plate. Various other tumours in the neck have been demonstrated with contrast enhancement.

The Spinal Cord

The use of contrast enhancement in the spinal cord with MRI differs quite considerably from that with myelography or CT myelography. With MRI there is no need for an intrathecal contrast agent in order to demonstrate the margins of the spinal cord, and intravenous Gd-DTPA can be used in a way that parallels its usage in the brain, i.e. to distinguish intramedullary and extramedullary tumours (Fig. 8.13) and to distinguish tumour from oedema (Fig. 8.14).

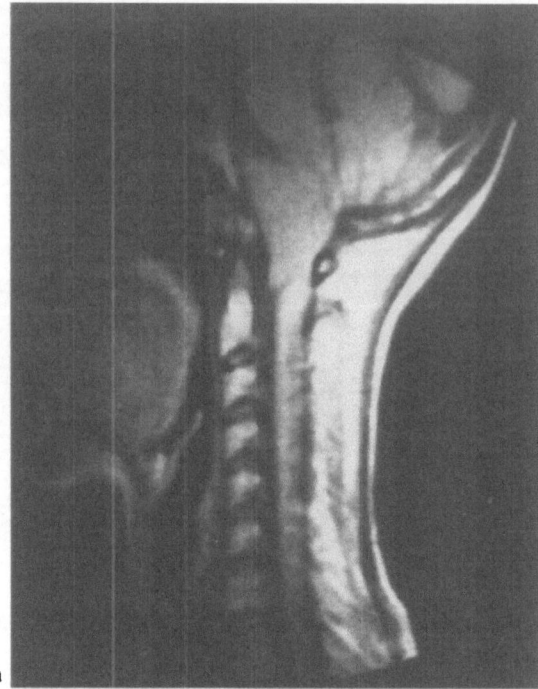

a

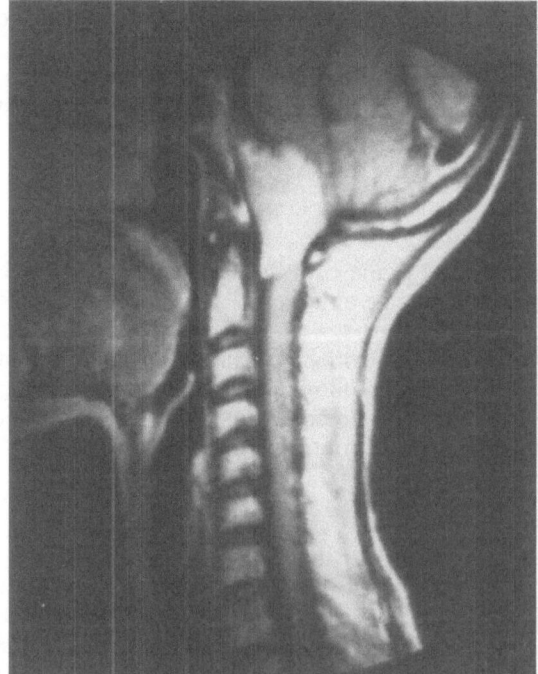

b

Fig. 8.13.   Meningioma: SE$_{544/44}$ scans **a** before Gd-DTPA and **b** after Gd-DTPA. The tumour at the foramen magnum displays moderate enhancement.

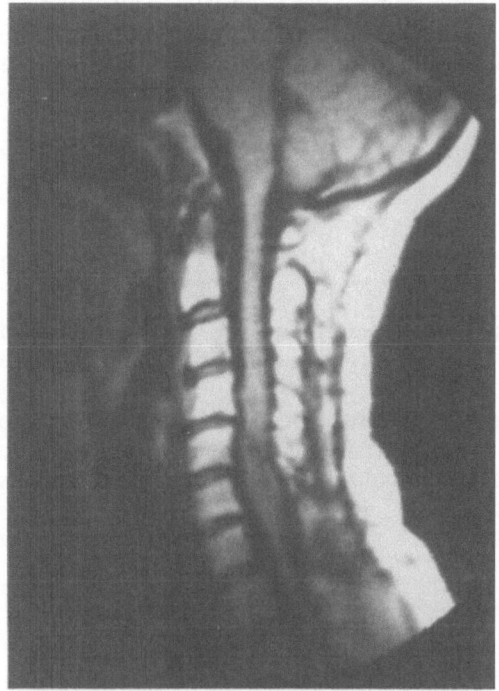

a

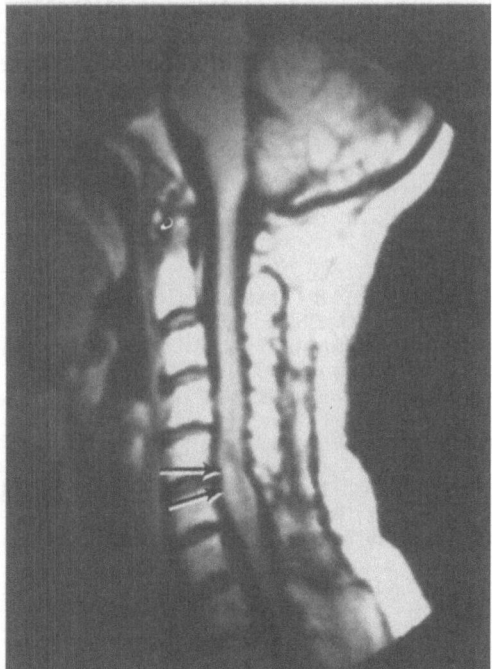

b

Fig. 8.14.   Ependymoma: SE$_{544/44}$ scans **a** before Gd-DTPA and **b** after Gd-DTPA. The tumour displays
a region of enhancement anteriorly (*arrows*).

In a series of 17 cases studied to date enhancement has been seen in all instances. The enhancement has been of value in distinguishing extramedullary tumours from subarachnoid cysts as well as defining the extent of intra-axial tumours. This appears likely to be an important application for Gd-DTPA and further studies are underway.

Toxicity

Patients have been questioned about symptoms before and after administration of Gd-DTPA. No significant side-effects have been reported and no significant signs have been observed.

Haematological and biochemical screening tests have been performed and to date the only significant finding of note has been a transient increase in serum iron in 10%–20% of cases. This has returned to normal within 24 h in all cases and its significance is uncertain at the present time.

Conclusion

Gd-DTPA appears to provide a valuable new approach to diagnosis in MRI. More studies will be required in order to define its role in clinical practice but there is no doubt that it is effective and in certain cases it can provide information not available with unenhanced MRI.

References

Bilaniuk LT, Zimmerman RA, Spagnoli M, Goldberg HI (1985) MRI of meningiomas. Proceedings of the 4th Meeting of the Society of Magnetic Resonance in Medicine. SMRM, Berkeley, CA, p 312

Bradley WG, Waluch V, Yadley RA, Wycoff RR (1984) Comparison of CT and MR in 400 patients with suspected disease of the brain and cervical spinal cord. Radiology 152: 695–702

Brasch RC (1983) Methods of contrast enhancement for NMR imaging and potential applications. Radiology 147: 781–788

Bydder GM, Kingsley DPE, Brown J, Niendorf HP, Young IR (1985) MRI of meningiomas (including studies with and without Gadolinium-DTPA) J Comput Assist Tomogr 9(4): 690–697

Bydder GM, Brown J, Niendorf HP, Young IR (to be published) Enhancement of cervical intraspinal tumours with intravenous gadolinium-DTPA. J Comput Assist Tomogr

Carr DH, Brown J, Bydder GM et al. (1984a) Intravenous chelated gadolinium as a contrast agent in NMR imaging of cerebral tumours. Lancet I: 484–486

Carr DH, Brown J, Bydder GM et al. (1984b) Gadolinium-DTPA as a contrast agent in MRI: initial clinical experience in 20 patients. AJR 143: 215–224

Claussen C, Laniado M, Kazner E, Schorner W, Felix R (1985) Application of contrast agents in CT and MRI (NMR): their potential in imaging of brain tumours. Neuroradiology 27: 164–171

Curati WL, Graif M, Kingsley DPE et al. (to be published) Magnetic resonance imaging: contrast enhancement of acoustic neuromas with intravenous gadolinium-DTPA. Radiology

Felix R, Laniado M, Claussen C, Schorner W, Weinmann HJ, Niendorf HP (1984a) Characterization of gadolinium-DTPA. Basic properties and first clinical results. 7th CARVAT, Rome, 6–10 Feb 1984

Felix R, Schorner W, Claussen C, Fiegler W, Niendorf HP (1984b) Diagnostic value of gadolinium-DTPA in MR imaging of brain tumours. Radiology 153(P): 84

Gadian DG, Payne JA, Bryant DJ, Young IR, Carr DH, Bydder GM (1985) Gadolinium-DTPA as a contrast agent in MRI — theoretical projections and practical observations. J Comput Assist Tomogr 9(2): 242–251

Graif M, Bydder GM, Steiner RE, Niendorf HP, Thomas DGT, Young IR (to be published) Contrast enhanced MRI of malignant brain tumours. AJNR and AJR

Koenig SH, Brown RD III (to be published) Relaxation of solvent protons by paramagnetic ions and its dependence on magnetic field and chemical environment: implications for NMR imaging. Magn Reson Med

Koenig SH, Baglin C, Brown RD III, Brewer CF (to be published) Magnetic field dependence of solvent proton relaxation induced by Gd^{3+} and Mn^{2+} complexes. Magn Reson Med

Leung AW-L, Steiner RE, Young IR (1984) NMR imaging of the liver in two cases of iron overload. J Comput Assist Tomogr 8(3): 446–449

McKay IM, Bydder GM, Young IR (in press) MRI of central nervous system tumours which do not display evidence of an increase in T_1 or T_2. J Comput Assist Tomogr

Runge VM, Clanton JA, Lukehart CM, Partain CL, James AE (1983) Paramagnetic agents for contrast enhanced NMR imaging: a review. AJR: 1209–1215

Schorner W, Felix R, Laniado M et al. (1984) Prüfung des kernspintomographischen Kontrastmittels Gadolinium-DTPA am Menschen. Fortschr Roengenstr 140: 495–500

Tripathi A, Bydder GM, Hughes JMB et al. (1984) Effect of oxygen tension on NMR spin-lattice relaxation rate of blood in vivo. Invest Radiol 19(3): 174–178

Weinmann H-J, Grier H (1983) Paramagnetic contrast media in NMR tomography — basic properties and experimental studies in animals. Proceedings of the 2nd Meeting of the Society of Magnetic Resonance in Medicine. SMRM, Berkeley, CA, pp 370–371

Weinmann H-J, Brasch RC, Press WR, Wesbey GE (1984) Characteristics of gadolinium-DTPA complex: a potential NMR contrast agent. AJR 142: 619–624

Zimmerman RD (1984) MRI in intracranial meningiomas. Proceedings of the 3rd Meeting of the Society of Magnetic Resonance in Medicine. SMRM, Berkeley, CA, pp 779–780

9. Tissue Characterisation by NMR

Margaret A. Foster

Introduction

Nuclear magnetic resonance has been used in the examination of tissues almost since the discovery of the phenomenon by Bloch and independently by Purcell in 1946. It is often asserted (but was not published) that the earliest NMR study in vivo was performed by Bloch who placed his finger into the NMR probe and obtained a proton signal from the digit. In vitro tissue studies were occasionally reported during the 1950s and 1960s, e.g. Odeblad et al. (1956), Bratton et al. (1965). It was, however, the reports of Damadian and his co-workers in the early 1970s (e.g. Damadian 1971), describing differences in proton relaxation times between normal and pathological tissues, which initiated the great surge of interest in the application of NMR to biological and medical studies. These studies have culminated in the widespread use of NMR spectroscopy and especially of imaging which we see today.

NMR is applicable to the study of a large number of nuclei, some of these, such as 31-phosphorus and 23-sodium, being of direct biological importance. The great majority of biological studies have, however, concentrated on the hydrogen nucleus, or proton, and the following discussion will be limited to proton NMR studies.

The nature of tissue and the methods used in biological NMR studies, both in vitro and in vivo, limit the chemical range of protons which contribute to the NMR signal. Protons attached to large molecules or to rigid or semi-rigid structures such as membranes do not contribute to the signal. In these cases the protons are very highly constrained and the T_2 relaxation time in extremely short. They are effectively "invisible" in normal tissue studies. If, however, the protons are part of rapidly-moving small molecules they maintain phase coherence after a 90° RF pulse for a much longer time and energy loss by spin-lattice interactions is also much slower. Hence, the protons of molecules such as water have T_1 and T_2 relaxation times in the measurable range. Apart from water, the only other protons which contribute significantly to tissue NMR signals are those of fat, i.e. triglycerides. In this case the long fatty acid $-CH_2$ chains have considerable mobility (unlike the chains of membrane lipids) and the rate of dipole and other interactions is reduced. Hence T_1 and T_2 are increased sufficiently to be measured, although T_1 is generally shorter in fatty than in "wet" tissue.

These limitations of measurement define the set of properties upon which we must base our attempts to achieve tissue characterisations in proton NMR studies. We can only observe water and triglycerides. We have only the standard NMR parameters, i.e. total NMR signal size (related to proton or spin density which is usually symbolised by the letters PD or the Greek letter ρ), the spin-lattice (T_1) and spin-spin (T_2) relaxation times and in certain experiments a small range of other properties including flow, diffusion and chemical shift. The first three of these characteristics have been most widely studied although the additional three are beginning to attract more attention as the limitations of using the three major (but interlinked) NMR parameters become increasingly obvious.

The NMR Parameters

The precise value for each of the NMR parameters in a particular sample is determined by the local environment experienced by the detectable protons in that sample. For most body tissues water protons are the major contributors to the NMR signal. We will therefore use water to illustrate the ways in which the values of the NMR parameters can be varied.

Proton Density and Signal Size

The NMR signal size is a difficult quantity to measure in absolute terms even in a small sample in an NMR spectrometer and it becomes extremely difficult to quantify in the NMR image. The signal size is related to the net magnetisation of the sample. When the proton spins are affected by the applied magnetic field they can go into one of two possible states of alignment — parallel or antiparallel to the direction of the applied field. There is a preferential distribution between these states, one of which is at a slightly higher energy level than the other, depending on the strength of the applied field and the sample temperature. This difference between the two populations yields the net magnetisation of the sample. As a working figure, the difference is about one in 10^6, i.e. only one proton in a million contributes to the NMR signal.

Temperature and applied magnetic field strength are fundamental factors affecting NMR signal size but there are many other factors, mainly instrumental, which also have an effect. In an "ideal" NMR experiment the signal would be measured as the free induction signal amplitude after delivering a perfect and infinitely short 90° RF pulse to a sample which had never previously experienced such a pulse, i.e. the sample was totally relaxed. In a real experiment, however, the 90° pulse is of finite duration, hence allowing some relaxation to take place during its delivery. Any imperfections, i.e. a larger or smaller pulse affecting the angle of the magnetisation, will affect signal size. In addition, the sample is usually subjected to a series of pulse sequences which at best are separated by fairly long inter-pulse intervals (t_r). If t_r is five times the T_1 relaxation time of the protons then the signal will be reduced to 99.3% of its theoretical maximum with further reductions in t_r causing proportionate diminution of the signal.

In the imaging situation the problems include all those discussed above for a simple sample. In addition there are further problems related to factors such as the necessity of using slice selection pulses which, by their nature, must be long, i.e. several milliseconds. Signal size is also affected by the use of field gradients, the greater tendencies for main field and RF inhomogeneity, etc. The proton or spin density signal displayed in an NMR image is not a direct measure of signal size — otherwise doubling the voxel size would double the proton density! It bears, however, a simple volume-scaled relationship to it.

Because of all the problems involved it is unwise to attempt any absolute measurement of proton density from an NMR image. The best that can usually be achieved is a relative measure obtained by placing a standard (often a mixture of H_2O/D_2O doped with copper or manganese ions to reduce T_1) in a position where it is visible in an appropriate region of the NMR image. Because of the extreme sensitivity of proton density to small variations in measurement conditions, errors can be introduced by examining the standard on different images (i.e. collecting a "phantom" image before or after the test image) or even by having the standard in a different region of the same image. It is also important to ensure that the standard is maintained at a fixed temperature.

Most NMR imaging pulse sequences in common use at the present time have a large component of proton density information in them. This may be weighted with a certain amount of T_1 and/or T_2 relaxation. Since proton density plays such an important part in determining the signal size obtained by sequences such as inversion recovery and saturation recovery it can readily be appreciated that any attempts at tissue characterisation based on values obtained from mixed parameter sequences are unlikely to succeed. To obtain pure proton (spin) density information presents problems enough.

Proton density alone yields a very low contrast NMR image, which is mainly a reflection of the small variation in water content between most body tissues. Table 9.1 lists T_1 relaxation time (measured in vitro at 2.5 MHz and room temperature) and water content (measured by drying to constant weight at 60°C) of a variety of tissues obtained from one adult male goat. The precision in both sets of measurements is in

Table 9.1. T_1 values and water content of fresh tissue samples from an adult male goat[a].

Tissue	T_1 (ms)	% H_2O
Blood serum	800	92.6
Bile	958	90.7
White brain (medulla)	315	73.7
White brain (cerebrum)	280	69.4
Grey brain (cerebrum)	412	85.2
Liver	140	71.6
Spleen	189	67.7
Testis	505	86.8
Kidney cortex	265	81.3
Kidney medulla	491	85.3
Perinephric fat	142	4.2
Pericardial fat	115	9.1
Myocardium	272	78.1
Thigh muscle	213	76.8

[a]T_1 values at 2.5 MHz and room temperature; water content by drying to constant weight.

the order of 7%. It can be seen that the low fat content tissues (i.e. "wet" tissues in this discussion) have a water content between 67.7% (spleen) and 86.8% (testis), i.e. a highest to lowest value ratio of 1.28 (fat is excluded from this since both lipid and water contribute to the NMR signal). The T_1 range of this same group of wet tissues, however, is 140 ms (liver) to 505 ms (testis), giving a ratio of 3.61. The difference between these ratios is an indication of the contrast available on an image produced using these tissue properties. Tissue water content is only an approximate guide to the observable proton density of wet tissues but the figures above serve to explain the very poor soft tissue contrast seen on pure proton density images. On such images the only distinguishable tissues are adipose (which has a higher proton density than wet tissues), bone (low proton density) and fluid pools such as CSF-filled ventricles of the brain, bile-filled gall bladder and urine-filled bladder.

T_1 and T_2 Relaxation Times

Tissue water does not exhibit the same relaxation characteristics as pure free water. If a sample of pure water is examined we find that $T_1 = T_2 =$ approximately 3 s and that this value is frequency independent. In tissue we usually find that T_1 is considerably longer than T_2 and that both of these are much shorter than the pure water values. T_1 also shows a marked frequency dependence. The differences in relaxation characteristics arise from constraints on the movement of water molecules in tissue due to the presence of the other constituents such as macromolecules, membranes, etc. There is considerable debate about the actual form of water within tissue but Fig. 9.1 attempts to present a general explanation. It shows the surface of a protein from which protrude the various amino acid residues. Some of these, such as -CH₃ are hydrophobic whilst others, e.g. ionic groups, have a strong attraction for water molecules. Some of the water is, therefore, held closely to the surface of the protein for a significant period of time. This is usually referred to as "bound" water.

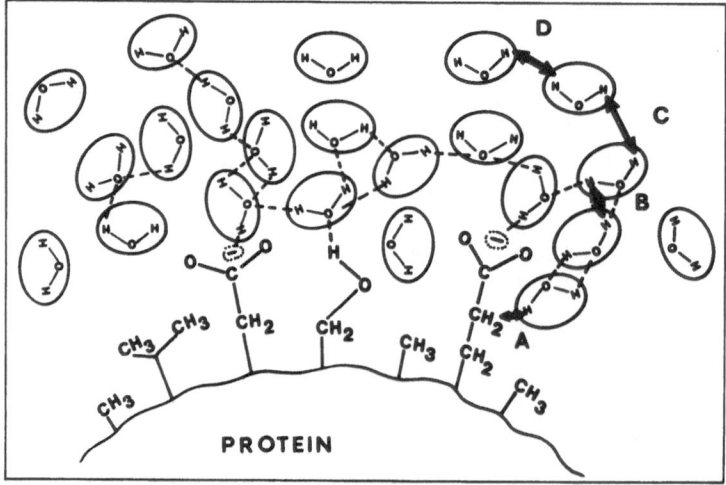

Fig. 9.1. General model of water association with the surface of a protein molecule.

Sometimes this binding is so close that spin exchange can occur with protons of the macromolecule (layer A); in other cases it is just a motional constraint (layer B). The attraction to the surface of the protein decreases with distance (layer C) until the constraint on movement of the water molecule is no greater than in the absence of the protein (layer D).

The presence of some constraint is universally accepted but the extent of this affect is still hotly debated. Suggestions range from a very few water molecules — possibly less than one complete layer over the surface of the macromolecules — to between 5% and 20% of the total cellular water. One theory even suggests that the presence of the macromolecular cell components structures the entirety of cellular water so that none of it has the freedom of movement of pure water. Limited constraint models are most widely accepted and account for the reductions in relaxation time on the basis of averaging of effects during the necessary measurement time for the NMR experiment. Tightly bound water undergoes extemely rapid dipole interactions and, if it could be measured alone, would show a very short (microsecond scale) T_1 value. Free water, as we have seen, relaxes in seconds. A single water molecule, during the time needed to measure its NMR relaxation rate, is likely to experience both of these environments since the "binding" is a very short period restraint. Hence the result of our NMR experiment is a relaxation time which is a weighted mean of all the environments which the water molecule has experienced during the measurement period. From this simplistic explanation it can be seen that if a large amount of water is just slightly constrained it will have the same effect on T_1 as if a small proportion of the water is very tightly bound. More detailed experimentation, including the use of methods other than NMR, tends to suggest that the latter case reflects the actual condition in living tissue.

A further complication in the measurement of tissue relaxation times is introduced by the fact that the necessary sample is of large size both for in vitro studies (generally in the order of 0.2–0.5 g) and in vivo where the "sample" is the contents of one image voxel. Hence we are not looking at water in a single cell or even, necessarily, of a single cell type. The measured value is therefore an average of all the T_1 or T_2 values from all the different environments inside and between the various cells and cell types which constitute the sample. Despite this apparent complexity of origin of the signal we find that in the majority of tissues the T_1 relaxation curve is of the simple, single exponential type that we would expect from the water protons in a protein solution. T_2 relaxation shows a greater tendency towards multi-exponentiality, suggesting that T_2 is a more sensitive indication of the complexity of the cellular environment.

Measurement Methods

Measurement of T_1 of tissue in vitro is most simply and most accurately performed by use of a simple inversion recovery sequence. In this case a 180° RF pulse is delivered to the sample to invert the spin population. A short period, t, is allowed for relaxation (spin-lattice or longitudinal relaxation only after a 180° pulse) and the sample is then subjected to a 90° RF pulse which puts the magnetisation into the xy plane in which it is measurable as the size of the free induction signal generated in a receiver coil. By varying the duration of t, one can obtain a plot of t against signal size. In a simple sample this plot will be a single exponential whose time constant is T_1, the spin-lattice relaxation time.

T_2 needs a more complex method of measurement and is normally obtained using some form of spin echo pulse sequence, e.g. that called the Carr-Purcell-Meiboom-Gill (CPMG) sequence (Farrar and Becker 1971). In this case a train of spin echoes is produced and the size of the echo is plotted against time after the initial 90° RF pulse. Again in a simple system this will yield a graph of single exponential form whose time constant is T_2.

It is seen, therefore, that normal in vitro experimental methods provide a decay curve with as many points on it as the experimenter chooses to obtain. From this a single value for the time constant can be obtained, e.g. by plotting the logarithm of the value and taking the slope of a calculated linear regression through the points, or the plot can be subjected to a more rigorous analysis if it is suspected to be multi-exponential. In NMR imaging, however, calculated T_1 and T_2 values are often taken from a very few points on the relaxation curve, making it very difficult or impossible to analyse for multi-exponentiality. Most relaxation time values quoted from NMR imaging must, therefore, be regarded as simplistic rather than accurate. For T_1, where most wet tissues show single exponential relaxation, this is less important than for T_2.

In imaging, T_1 is normally obtained by comparison of signal sizes using interleaved saturation recovery and inversion recovery pulse sequences. In the first of these the signal, in theory, contains only proton density information, i.e. is not affected by relaxation. The inversion recovery signal is T_1 weighted to an amount defined by the relationship between t_i (the interval between inversion and the 90° pulse — equivalent to the "t" interval for in vitro T_1 measurements) and T_1. If the inter-pulse interval (t_r) is sufficiently long that no saturation of either signal occurs then those two will provide the size of initial magnetisation (from the SR sequence) and a single point on the relaxation curve (from the IR sequence). Again assuming that T_1 is a single exponential phenomenon it is possible from these two values to produce a reasonable spin-lattice relaxation time. The method is, however, prone to inaccuracies if t_r is too short or if the relaxation is multi-exponential.

T_2 is calculated from a multiple spin echo sequence in much the same way as was discussed for CPMG earlier. Once again the use of a short t_r or looking at a tissue with multi-exponential relaxation will upset the results.

In addition to inaccuracies generated by the pulse sequence, other aspects of NMR image collection can affect T_1 and T_2 values calculated from image data. Two examples of potential problems, which are not necessarily independent, are peculiarities of the slice profile and partial voluming effects. The slice profile refers to the shape of the slice through its width, that is the NMR sensitivity at right angles to the image plane. The slice profile is defined by the relationship between the field gradient applied during selective excitation and the shape of the slice selective RF pulse (usually 90° pulse, although selective 180° pulses have been examined). Figure 9.2 shows typical slice profiles from three types of RF pulses which can be used in NMR imaging. It can be seen that the sensitivity across these is very different and hence, if used to image inhomogeneous material (e.g. the brain where grey and white matter are intimately mixed) the resultant relaxation time values will be different since the image voxel will contain different proportions of grey and white matter and the relaxation time will be an average of the contribution of these tissues.

This leads on to partial volume effects. Even if a perfectly rectangular slice profile could be obtained, the voxel would in many cases still contain several different tissues — known as partial voluming. This will lead to an inaccurate value for relaxation time if this is calculated on the basis of a two-point plot as described above, or by use

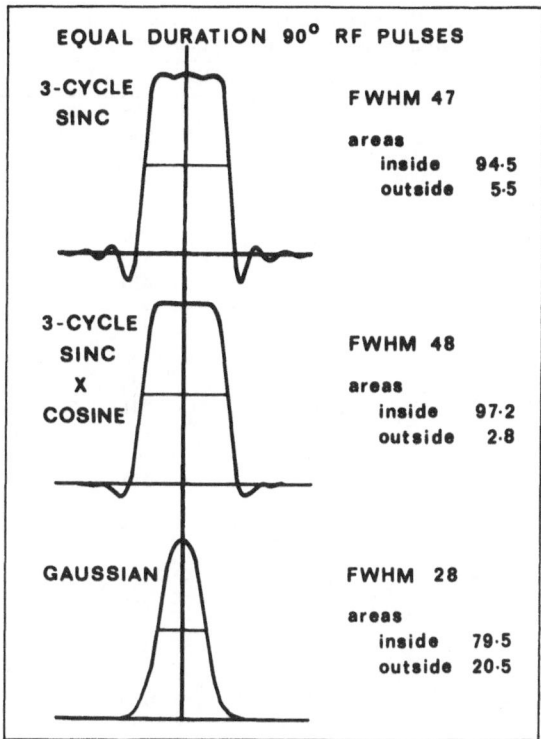

Fig. 9.2. Slice profiles derived from three possible RF pulse shapes. FWHM is full width at half maximum height of profile.

of a linear regression through a very few points. If a "full plot" method is used to produce a full relaxation curve, then the presence of more than one tissue will produce multi-exponentiality in the relaxation curve. This can lead to extreme complexity, especially in T_2, if the relaxation characteristics of any of the tissues in the voxel are individually multi-exponential. Analysis of such complexity becomes impossible.

In summary, therefore, we can see that relaxation times reported from NMR images must be treated with caution. In general these are acceptable within the context of a specified pulse sequence on a particular instrument but the values are rarely "absolute" and intercomparison between instruments is hazardous. It should also be remembered that the field strength (or operating frequency) affects the T_1 value (Farrar and Becker 1971) although this is much less important for T_2 values.

Range of T_1 and T_2 values

A few values for T_1 and T_2 of normal human tissues, measured in vivo, are given in Table 9.2. These have been obtained from the literature as reviewed in Bottomley et al. (1984) and due to the differences in measurement methods and associated problems discussed above, they should be taken only as a rough guide. The general trend is for T_1 value to increase with increasing measurement frequency whereas T_2 is less affected. The frequency dependence of T_1 value is demonstrated in Fig. 9.3 in which rat tissues are examined at constant temperature (30°C) and various

Table. 9.2.   Relaxation values for some normal human tissues measured in vivo (data from review by Bottomley et al. 1984)

Tissue	Frequency (MHz)	T_1 (ms)	T_2 (ms)
Liver	1.7	155	–
	6.5	250	–
	8.5	380	40
	15.0	–	49
Spleen	1.7	270	–
	6.5	510	–
	8.5	420	20
	15.0	–	61
Muscle	1.7	130	–
	8.6	400	50
	15.0	541	35
Grey brain	1.7	275	–
	6.5	515	–
	6.5	–	110
	12.8	600	100
White brain	1.7	225	–
	6.5	285	–
	6.5	–	105
	12.8	380	80
Adipose	1.7	130	–
	8.5	240	60
	15.0	218	61

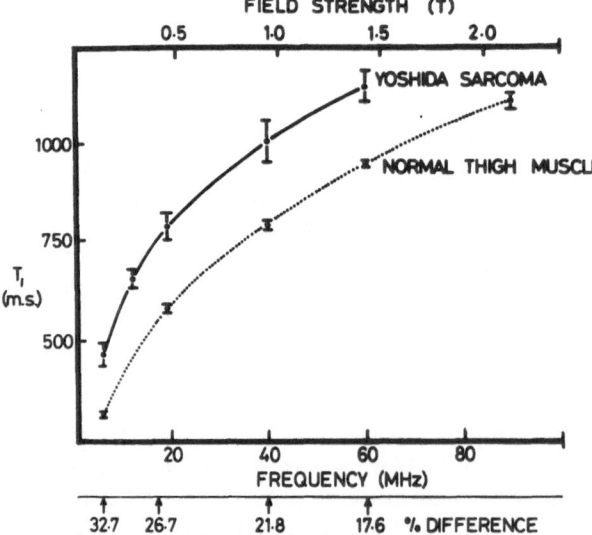

Fig. 9.3. T_1 values of Yoshida sarcoma and normal thigh muscle from rats, examined at 30°C and variable NMR frequency. Note the line below the graph indicating the difference in frequency dependence between the tissues.

frequencies. The tissues are normal thigh muscle from control rats and samples of Yoshida sarcoma grown subcutaneously in the rat thigh. The figure demonstrates the increase in T_1 with frequency shown by both tissues, and particularly the difference in frequency dependence exhibited by the two tissues. Figure 9.4 demonstrates the same phenomenon (frequency here is on a logarithmic scale). In this case, the values have been obtained from the published literature of 26 different groups of workers. They are selected for a measurement temperature range between 24° and 30°C. It can be seen that normal muscle and liver (from a variety of mammalian species) have different frequency dependences. In contrast, Fig. 9.5 shows a lack of frequency dependence of T_2 values for liver and spleen tissue.

Physiological or biological factors can also affect the relaxation times of tissues. Figure 9.6 shows calculated T_1 images through the thighs of a normal volunteer (a) after resting and (b) after taking intensive exercise (running up and down stairs). The T_1 value of the exercised part of the muscle is considerably elevated in (b), presumably due to the increased blood supply within the muscle.

Normal development is also associated with changes in relaxation behaviour of the maturing tissues. The foetus and neonate have a very high water content in almost all body tissues. The T_1 value of the tissues is also higher than that seen in comparable adult organs. As the animal matures, both water content and T_1 fall as illustrated for certain rat tissues in Fig. 9.7. Rate and extent of maturation, however, vary between the tissues. Spleen shows very little effect at all, whereas thigh muscle T_1 falls to less than 40% of the neonate value over a period of 4–5 weeks. Most tissues, e.g. liver and brain tissue, have reached the adult value sooner than has the voluntary muscle.

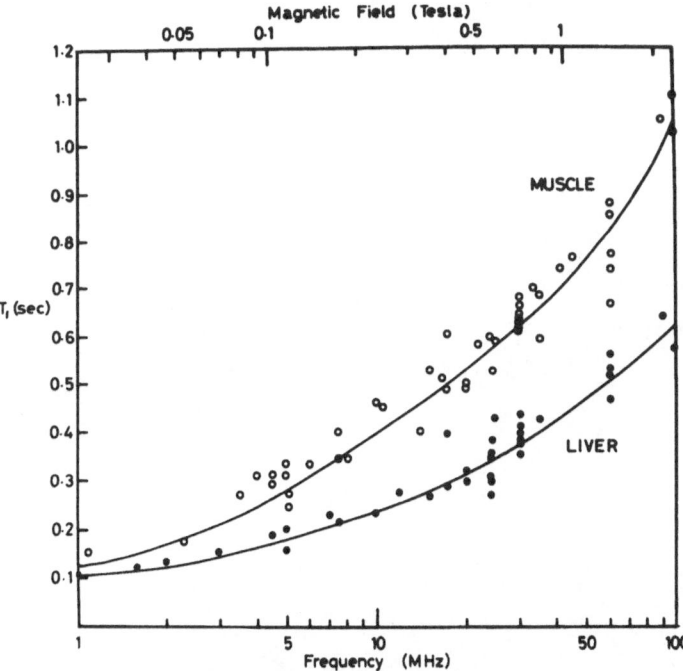

Fig. 9.4. T_1 value compared to measurement frequency of muscle and liver. Results from 26 groups obtained between 24°C and 30°C using a variety of species.

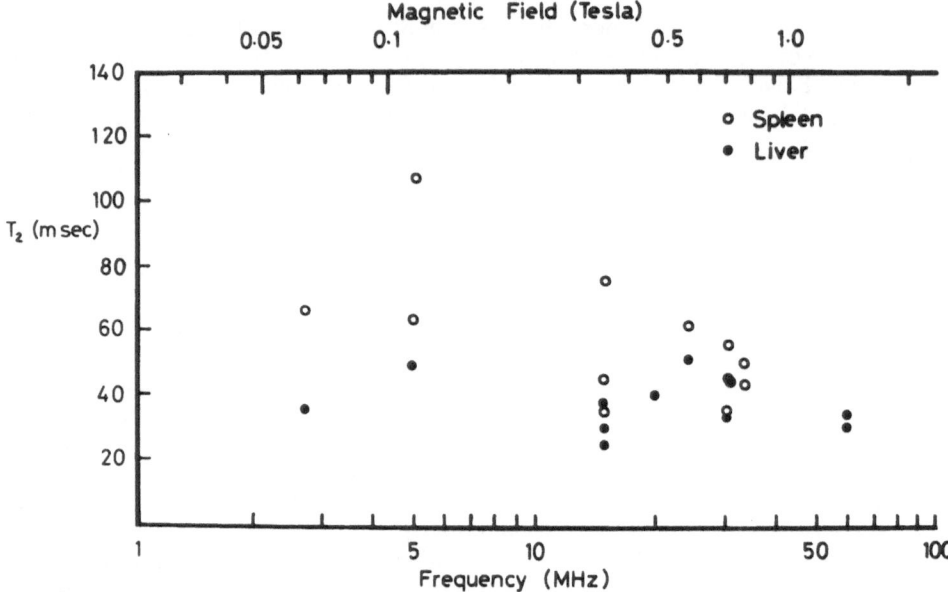

Fig. 9.5. T2 value compared to measurement frequency of spleen and liver. Results from eight groups obtained between 18°C and 30°C using a variety of species.

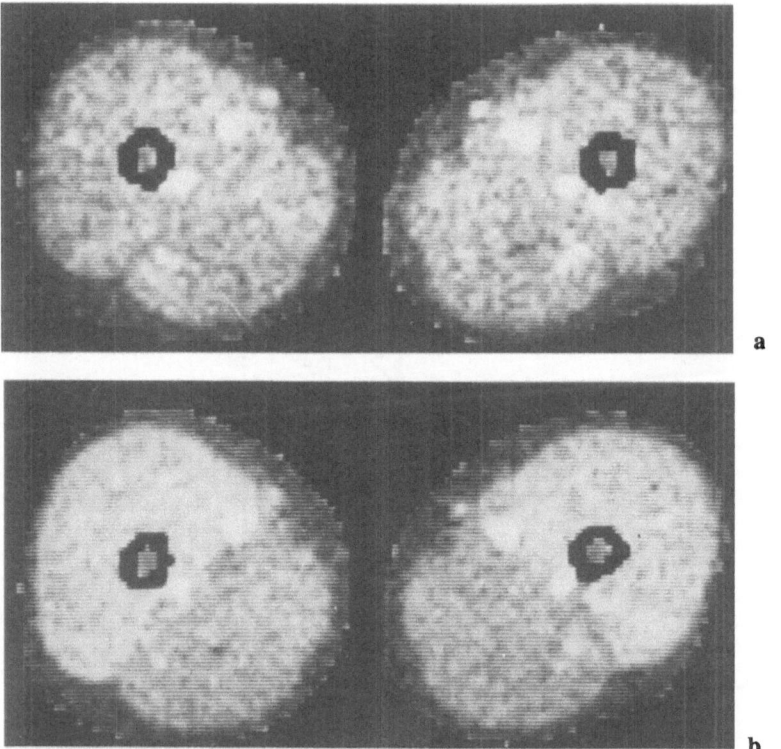

Fig. 9.6. Calculated T_1 images through thigh region of a normal volunteer, **a** before and **b** after exercise. Increase in brightness indicates a longer T_1 value. (Image obtained at 1.7 MHz.)

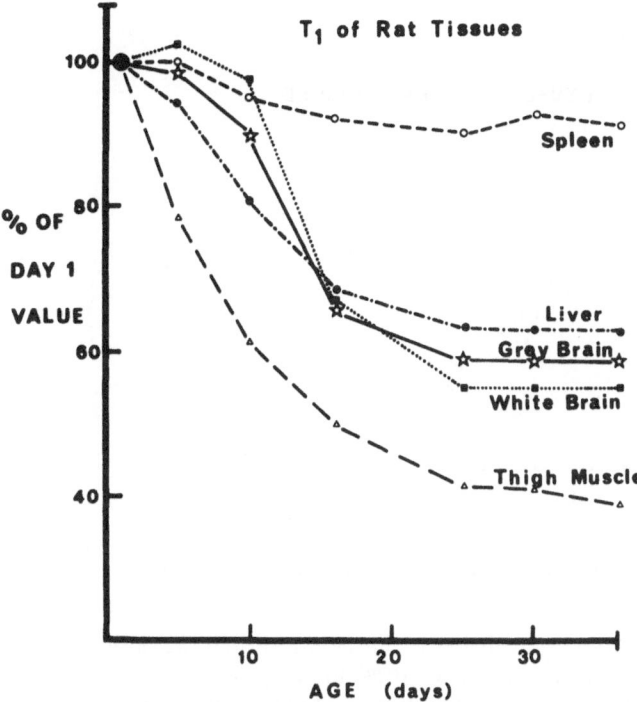

Fig. 9.7. Variation of T_1 value of a variety of immature rat tissues compared with age of the animal. The T_1 is expressed as percentage of the value obtained on the first day of neonatal life. Measurements at room temperature and 2.5 MHz.

White brain shows a very odd pattern in that it stays at neonate T_1 value for about 10 days then, co-incident with the onset of myelination, it falls extremely rapidly. From this it can be seen that NMR imaging studies of young individuals have to be undertaken with caution and with full knowledge of the different maturation characteristics of the tissues. Otherwise, the images are open to considerable misinterpretation. The time scale in the human is much longer, myelination, for example, taking 3 years or more to approach completion.

Although water content and T_1 both fall during maturation they are not necessarily directly linked. If T_1 is a simple reflection of water content then a plot of the relaxation rate (R_1 which is $1/T_1$) against water content would be linear. This is the course for some tissues such as developing liver. For muscle, however, it is not so, as is shown in Fig. 9.8. Changes in the macromolecular composition and/or paramagnetic ion content of the tissue could cause an effect of this type.

Possibilities for Tissue Characterisation

So far we have looked at the problems involved in trying to obtain meaningful values for the three major NMR parameters. Let us now briefly consider why one should attempt to obtain these separate values.

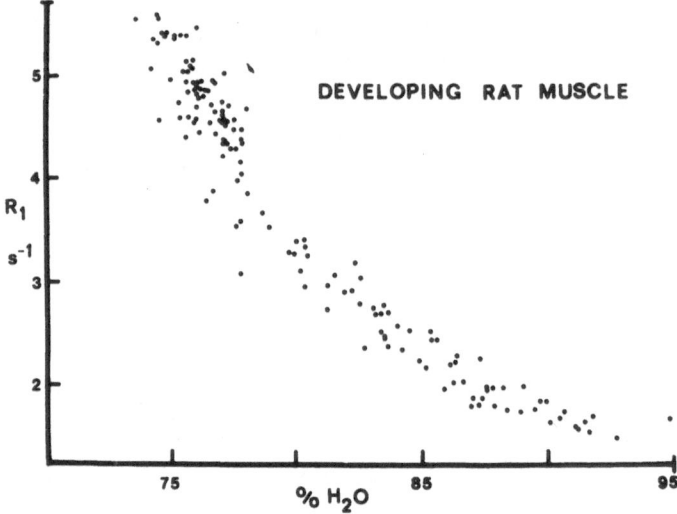

Fig. 9.8. Spin-lattice relaxation rate of developing rat thigh muscle compared to the tissue water content measured on the same sample. (NMR measurement at room temperature and 2.5 MHz).

Firstly, an objective which is not truly tissue characterisation is the optimisation of pulse sequences for NMR imaging. Tissue contrast in the NMR image is based on a complex relationship between the order and timing of events in the pulse sequence and the proton density, T_1 and T_2 values of the tissues in the selected image slice. It is normal practice in most clinical NMR departments to establish a small number of standard pulse sequences which, by experience, are known to demonstrate contrast between normal tissue and most pathologies. This, however, can be a dangerous procedure in that false negatives can be produced simply because an inappropriate pulse sequence was used and no abnormal/normal tissue contrast was achieved. On the other hand, to try every possible pulse sequence would not be possible. If, however, values for PD, T_1 and T_2 can be obtained then either an optimum pulse sequence can be devised and used or, in fact, the patient can be sent away since from these fundamental values any type of proton NMR image can be re-created (with the exception of diffusion or chemical shift images). The use of derived images from calculated parameter values is, perhaps, one direction for future development of NMR imaging.

A second reason for obtaining parameter values is to assist in true tissue characterisation, i.e. to obtain a unique identification of a particular normal or abnormal tissue based on its NMR characteristics. Let us examine the possibilities for achieving this unique distinction of the tissues, firstly, for normal tissues. The values quoted in the tissues so far have been either from a single individual (Table 9.1) or averages with, deliberately, no indication of scatter of results. For comparison Fig. 9.9 shows the scatter of results for a variety of tissues from similar animals. These tissue samples were all from young adult female Sprague Dawley rats which were killed on the 36th day after giving birth at the end of their first pregnancy. The samples were all handled in the same manner and T_1 value was measured at 2.5 MHz at a room temperature of 20°C ± 4°C (Fig. 9.9a). Water content measurement was achieved by drying out the samples used for NMR analysis, to constant weight at 60°C (Fig. 9.9b). In other words, scatter of results due to experimental conditions

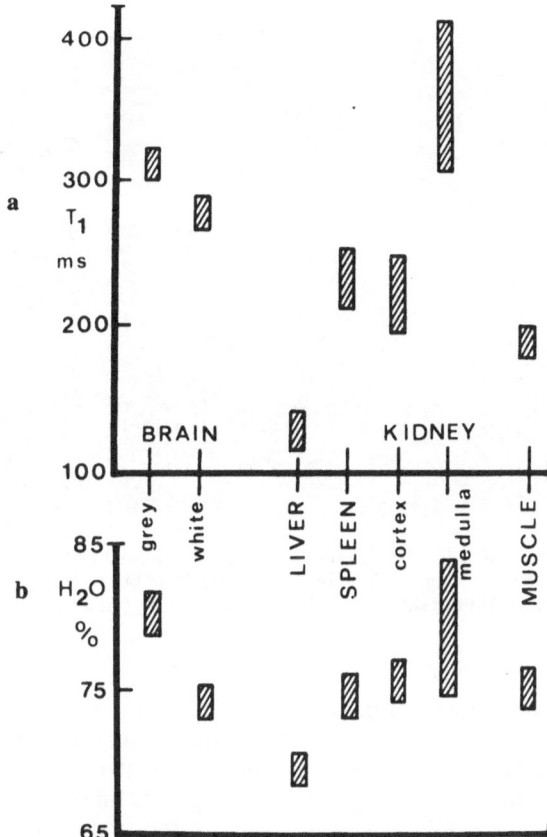

Fig. 9.9 Scatter of values for **a** T_1 (room temperature and 2.5 MHz) and **b** percent water content of normal rat tissues.

was as small as possible and we are observing the normal "biological variations" of the tissue itself. It is seen that the normal variation is quite large and, of greater importance for characterisation, this variation results in considerable overlap between tissue types. This is particularly the case for water content where a good separation is obtained for white and grey brain tissue and liver is also very distinctive. It would not, however, be possible to separate spleen, kidney cortex or medulla or voluntary muscle on the basis of water content. T_1 values show a little improvement on this, with kidney medulla now being separate from cortex and voluntary muscle showing no overlap with spleen.

The difference in proton density between white and grey brain tissue is surprising in view of the relatively small difference in proton density which is observed on the NMR image. Compare Fig. 9.10a, which shows a proton density, i.e. saturation recovery, sagittal image through the normal head with Fig. 9.10b, which is a calculated T_1 coronal image from the same individual. This unexpectedly low proton density contrast may, in part, be due to saturation effects since other workers, e.g. Wehrli et al. (1984) have found larger differences in proton density using very long saturation recovery sequences. Wehrli reported relative proton densities of 0.72 for white matter and 0.84 for grey matter compared with 1.00 for CSF.

In Fig. 9.11a the values given in Fig. 9.9 are plotted individually to show the grouping in two dimensions with T_1 and water content as the dimensions. In this case,

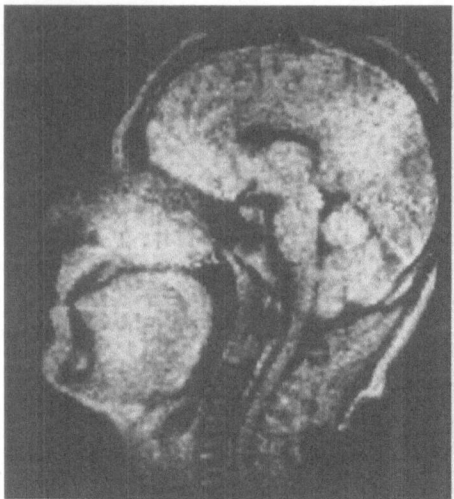 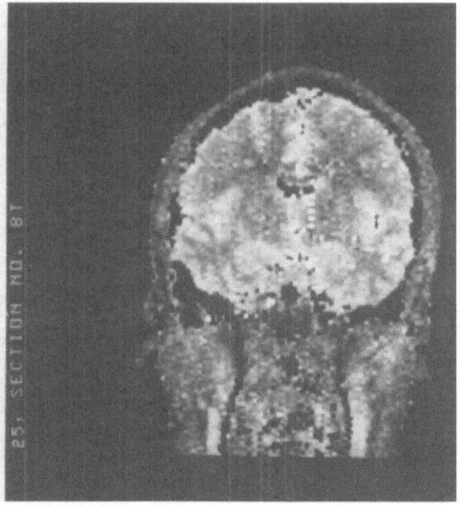

a   b

Fig. 9.10a, b. NMR images through the normal head obtained at 3.4 MHz **a** Proton density sagittal image showing considerable anatomical detail but poor grey/white tissue discrimination. **b** Calculated T_1 coronal image showing poorer general anatomy but improved contrast of brain tissue.

where two parameters are considered for each tissue, there is a slightly better separation with each tissue forming a unique "island" on the diagram. There is still, however, too close a relationship between spleen, kidney cortex and muscle for certainty of identification based purely on these values.

Figure 9.11b takes this a stage further and attempts to include T_2 as a third dimension. Unfortunately, the experiment quoted in the earlier plots did not include obtaining T_2 values. To obtain some T_2 information, therefore, all the T_2 values for each tissue (as quoted in the Bottomley review) were averaged. The values were selected only on the basis that they were obtained at temperatures between 15°C and 25°C — comparable to the actual experiment quoted here. In addition T_2 values were only included if the authors had quoted a single value, i.e., had made no attempt to separate out components of possible multi-exponential relaxation curves. On this basis an average T_2 was obtained for each tissue (except kidney medulla — most of the quoted T_2 values were for kidney as a whole and were attributed mainly to kidney cortex). The tissue "islands" were then shifted at 45° on the plot, in accordance with the T_2 axis.

It can be seen that the spleen, having a slightly longer T_2 than kidney, is now a little better separated from it although the usefulness of this is very debatable. Voluntary muscle, however, has a longer T_2 than kidney and is now seen to overlap this tissue (and it would on a truly three-dimensional display of this plot). In this particular case, therefore, we have not improved tissue discrimination by using all three of the NMR parameters. The crudity of the T_2 values must, however, be taken into consideration. It is also necessary to consider which tissues are likely to be confused. White and grey brain tissue are intimately intermixed and it is of primary importance to be able to distinguish between them, and also between solid brain tissue and CSF. This is, as we have seen, relatively easy. Spleen and muscle are unlikely to be confused and in any case the T_1 image (Fig. 9.12) clearly distinguishes spleen from back muscle (the

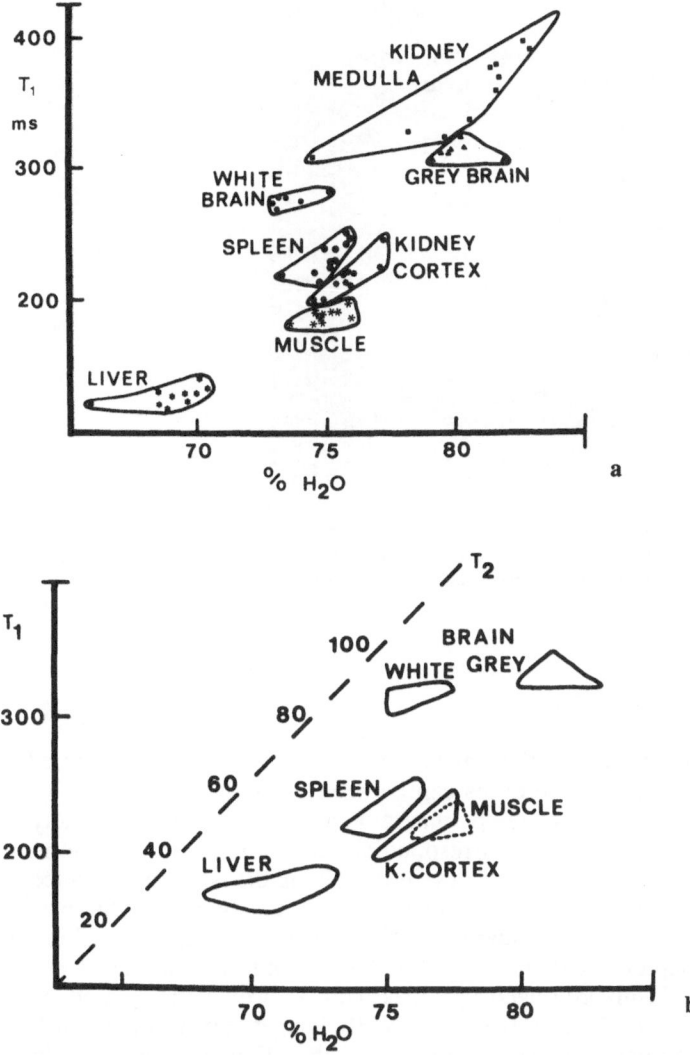

Fig. 9.11. Values for rat tissues used to produce Fig. 9.9 are plotted as **a** T_1 value against water content of the tissue sample and **b**, as in **a** but with a third dimension of T_2 weighting.

muscle used in the experiment above was from the thigh). The only major problem, among the tissues described above, is likely to be separation of spleen and kidney cortex. These can appear on the same section and are not clearly distinguished although the presence of perinephric fat frequently helps to separate the organs.

It appears likely, therefore, that the use of anatomical information along with all three major NMR parameters can uniquely distinguish a large number of normal tissues in the body. This presumably could be done on an automatic basis and, indeed, computer characterisation of brain tissue, using the three parameters, has already been successfully achieved, for example by Bachus et al. (1985). Similar methods to those described above are being investigated by Bielke et al. (1985).

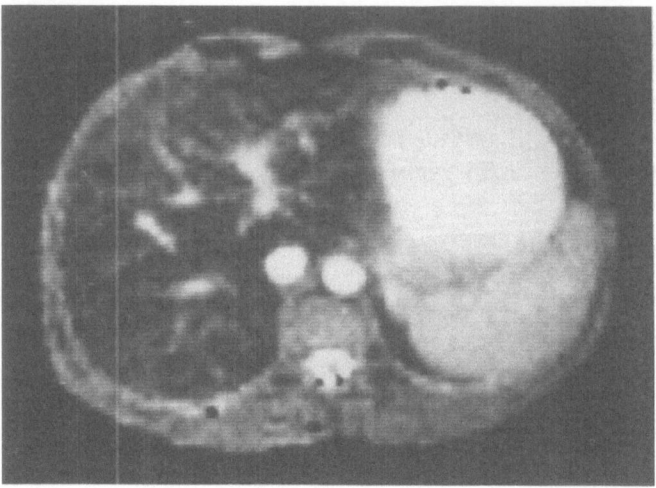

Fig. 9.12. Calculated T_1 NMR image (3.4 MHz) through the normal upper abdomen. The *dark region* represents the low T_1 value of liver. The *white region* is a fluid-filled stomach with the grey (intermediate T_1 value) spleen below this. Note contrast between spleen and body wall muscle.

Characterising Abnormal Tissues

Possibilities for characterisation of abnormal tissues are not so optimistic as those for normal tissues. Considerable differences can be seen on both calculated T_1 (and sometimes on T_2) images and on mixed parameter images between a huge variety of abnormalities and the surrounding or corresponding normal tissues. At the moment several hundred papers are published each year describing the NMR appearance of different diseases and conditions. The problem is to try to obtain quantitative, as opposed to qualitative, information about the abnormalities. As mentioned in the introduction to this chapter, the foundation of our current interest was the discovery that cancer is associated with prolonged T_1 values. This was well demonstrated in the early reports and can often be clearly seen in NMR images. For example, in Fig. 9.13a we see a proton density image and in Fig. 9.13b a T_1 image through a liver containing metastatic tumour nodules. These are poorly, if at all, displayed on the PD image but are very well shown on the T_1 image. Figure 9.14 shows a T_1 image of a brain glioma which also has a prolonged T_1 value and hence is observable. The mass of information available at this time, however, has shown us that not all tumours are so characteristic, e.g. meningioma in the brain frequently has a T_1 value which is indistinguishable from that of cortical grey matter. Also, even for tumours which can usually be seen, there tends to be a wide range of T_1 and T_2 values and these frequently overlap with the prolonged relaxation times associated with other conditions such as oedema. The use of contrast agents can be of great assistance in some of these cases and indeed some attempts have been made at localisation of tumours by means of calculated NMR parameters. For example, the work of Bachus showed that a glioma, the surrounding oedema and the normal brain tissue were separable and uniquely identified by a combination of T_1, T_2 and PD. In this case, however, the program needed to be "told" the values and these were obtained from

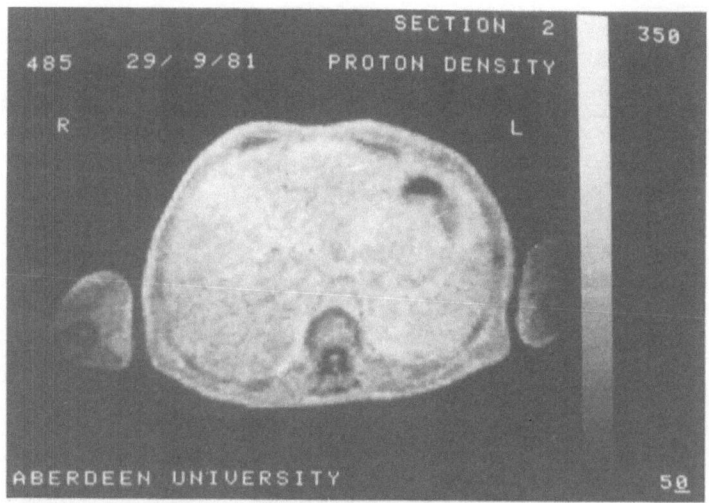

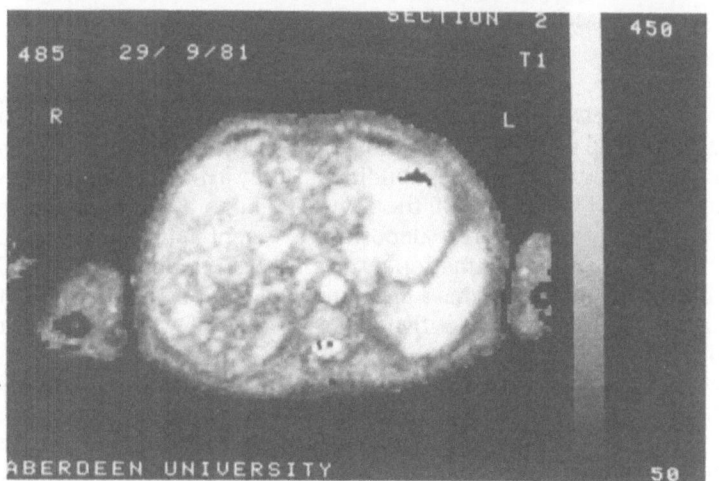

Fig. 9.13a, b. NMR images of upper abdomen with metastatic tumour nodules (3.4 MHz). **a** Proton density image showing generally poor soft tissue contrast; **b** T_1 image clearly displaying the nodules.

the lesion itself. This is rather the reverse of tissue characterisation in which one would hope to have a "look-up table" of parameter values from different tissues and abnormalities with some kind of probability rating. Given the values from the lesion one would then compare it with the tables and assign it to a tumour-type or other abnormality, within the limits of the defined probability. This, however, is not possible on the basis of the three major NMR parameters. The range of values for each abnormality tends to be very large, whilst the differences between many abnormalities is relatively small. Overlap is therefore too great to allow characterisation of this type. This has been shown (usually for a single parameter) by a number of workers who have examined parenchymal liver disease and shown that tendencies for an increase in T_1 occur in cirrhosis, inflammatory conditions, etc.

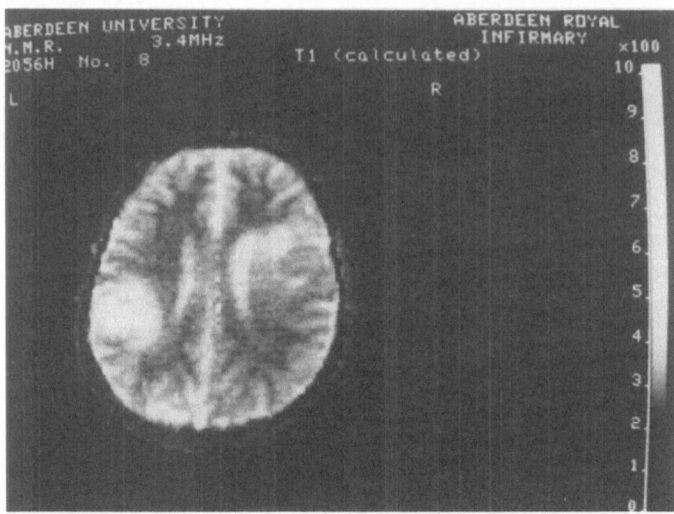

Fig. 9.14. Calculated T_1 image through brain showing tumour mass in one hemisphere.

There is, however, overlap between the values in the various disease groups and in most cases there is also overlap with the normal range. Hence the T_1 value is not, on its own, able to characterise any of the disease conditions. Only in haemachromatosis is there a significant trend. This is one of the few conditions associated with a reduction in liver T_1 value (due to iron deposition). Even here, however, a combination of iron build-up and inflammation or cirrhosis can occur which would tend to put the T_1 value back into the normal range, and could give a false negative if the T_1 was used in an attempt to characterise the disease. T_2 and PD values were not reported in this study.

The prospects for tissue characterisation on the basis of the three major NMR parameters appear, therefore, to be gloomy as far as abnormal conditions are concerned. This may, however, be too general a statement: there are diseases where a combination of the appearance of the condition and the NMR parameters is quite unique. For example, in multiple sclerosis the demyelination plaques frequently occur close to the ventricles. These regions have a T_1 value similar to the CSF and on T_1 images it is difficult to distinguish the plaques other than by their shape (Fig. 9.15a). If, however, another parameter is added, such as PD by using an inversion recovery image, then the plaques become clearly distinguished (Fig. 9.15b) since their proton density is less than that of the CSF and hence there is less saturation of the signal, making the plaque appear brighter (image at 3.4 MHz using IP 1000/200).

As well as looking for tissues which are well characterised, we must also consider the fact that no medical diagnosis is ever made on the basis of one test. For many conditions the NMR changes, taken alone, are non-specific. If, however, they are taken in the context of the full battery of tests to which the patient is subjected it may well be that the NMR parameters become diagnostically significant when other conditions of similar NMR characteristics but different medical features are ruled out.

The major use of NMR imaging at the moment is for anatomical display of the position and extent of lesions. The possibilities, at this time, for tissue

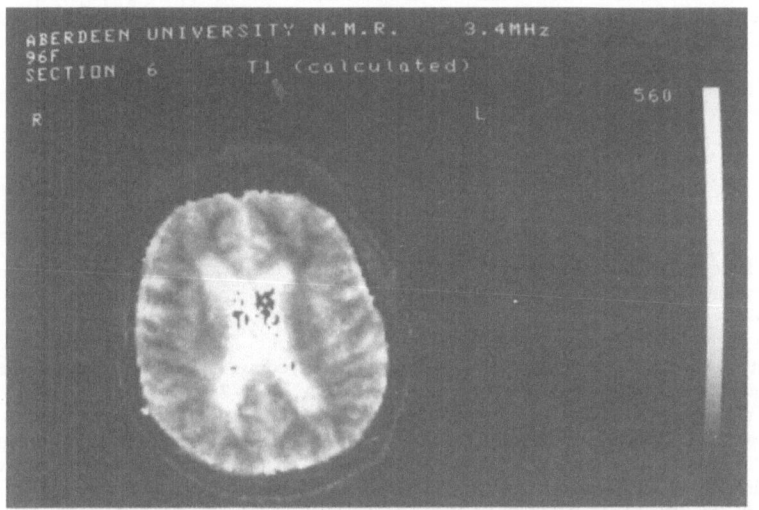

a

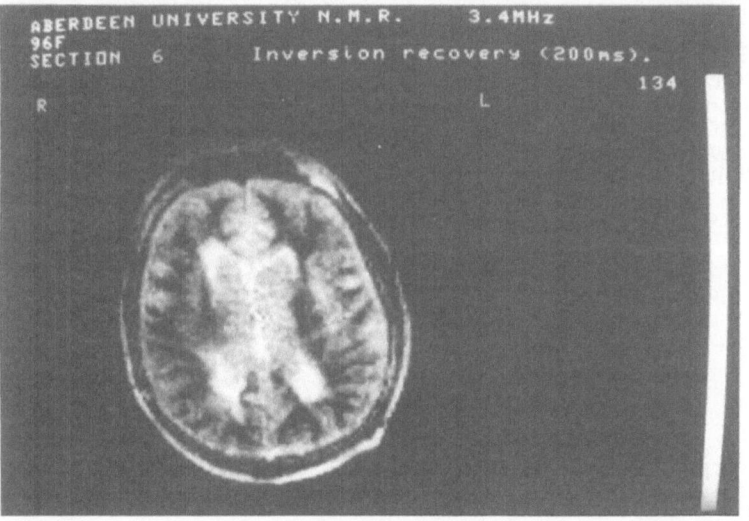

b

Fig. 9.15a, b. NMR images through brain of patient with multiple sclerosis. **a** T_1 image gives poor distinction between the sclerotic plaques and the ventricles; **b** IR 1000/200 image combines T_1 and PD information to make a clear demonstration of the lesions.

characterisation are limited for many diseases but medical NMR is still in its infancy and the underlying biophysical reasons for the changes in NMR appearance are not yet understood. It is, therefore, too early to limit the possibilities for NMR. Over the next few years we must continue to improve the quality of NMR images, to explore all the new possibilities as they become available (chemical shift imaging, diffusion imaging, etc.) and especially to deepen our understanding of the basic phenomenon. Once we truly understand what tissue features control the NMR characteristics, perhaps we will be able to achieve much more specific NMR tissue characterisation.

Acknowledgements. The author would like to thank Prof. J.R. Mallard and her other colleagues in the NMR teams for their help and encouragement. The work was, in part, supported by funding from the Medical Research Council, grant numbers SPG 7509376 and 98126010.

References

Bachus R, Koenig H, Lenz G, Deimling M, Reinhardt ER (1985) Tissue differentiation in MRI by means of pattern recognition. Proceedings of the 4th Meeting of the Society of Magnetic Resonance in Medicine. SMRM, Berkeley, CA, pp 18–19

Bielke G, Bruckner A, Meindl S, Seelen van W, Higer HP, Pfannenstiel P, Meves M (1985) A method for a multiparametric tissue characterisation in NMR-imaging. Proceedings of the 4th Meeting of the Society of Magnetic Resonance in Medicine. SMRM, Berkeley, CA, pp 22–23

Bloch F, Hansen WW, Packard M (1946) The nuclear induction experiment. Phys Review 70: 474–485

Bottomley PA, Foster TH, Argersinger RE, Pfiefer LM (1974) A review of normal tissue hydrogen NMR relaxation times and relaxation mechanisms from 1–100 MHz: Dependence on tissue type, NMR frequency, temperature, species, excision and age. Med Phys 11: 425–448

Bratton CB, Hopkins AL, Weinberg JW (1965) Nuclear magnetic resonance studies of living muscle. Science 147: 738–741

Damadian R (1971) Tumour detection by nuclear magnetic resonance. Science 171: 1151–1153

Farrar TC, Becker ED (1971) Pulse and Fourier transform NMR. Academic Press, New York

Odeblad E, Bahr BN, Lindstrom E (1956) Proton magnetic resonance of human red blood cells in heavy water exchange experiments. Arch Biochem Biophys 63: 221–225

Purcell EM, Torrey HC, Pound RV (1946) Resonance absorption by nuclear magnetic moments in a solid. Phys Review 69: 37–43

Wehrli FW, MacFall JR, Glover GH, Grigsby N (1984) The dependence of nuclear magnetic resonance image contrast on intrinsic and pulse sequence timing parameters. Magnetic Resonance Imaging 2: 3–16

10. The Influence of NMR Parameters on Imaging

P. Pfannenstiel, G. Bielke, S. Meindl, W. von Seelen
and H.P. Higer

Introduction

During the past 2 years we have conducted a study supported by the Federal Ministry of Research and Technology to examine the clinical correlation of NMR as well as the influence of NMR parameters on imaging using a prototype of a Bruker air coil whole-body magnet system with a field strength of 0.14 Tesla.

Biological tissues can differ mainly in four NMR-relevant characteristics: proton density, the relaxation times T_1 and T_2, and the blood flow velocity in the examined tissue. Since imaging and tissue differentiation are considered to be the potential values of NMR, diagnostic judgement should take into account all measurable NMR parameters — depending on the clinical problem.

Qualitative Evaluation with Appropriate Pulse Sequences

Three different procedures allow the evaluation and interpretation of NMR tomograms:

1. Differentiation of contrast shadings in a grey scale.
2. Discrimination as a result of various measurement modalities such as saturation recovery, inversion recovery and spin echo technique as well as changing of the corresponding parameters TE, TR and TI.
3. Quantitative assessment of proton density ρ, the spin-lattice T_1, spin-spin relaxation parameters T_2 and flow phenomena.

In addition to the intrinsic parameters shown in the upper part of Fig. 10.1, numerous extrinsic parameters (Bielke 1984), as seen in the lower part of Fig. 10.1, influence the generation of NMR images. These are field strength, radio frequency (RF field) and field gradients. Thickness, distance and shape of the slices selected for

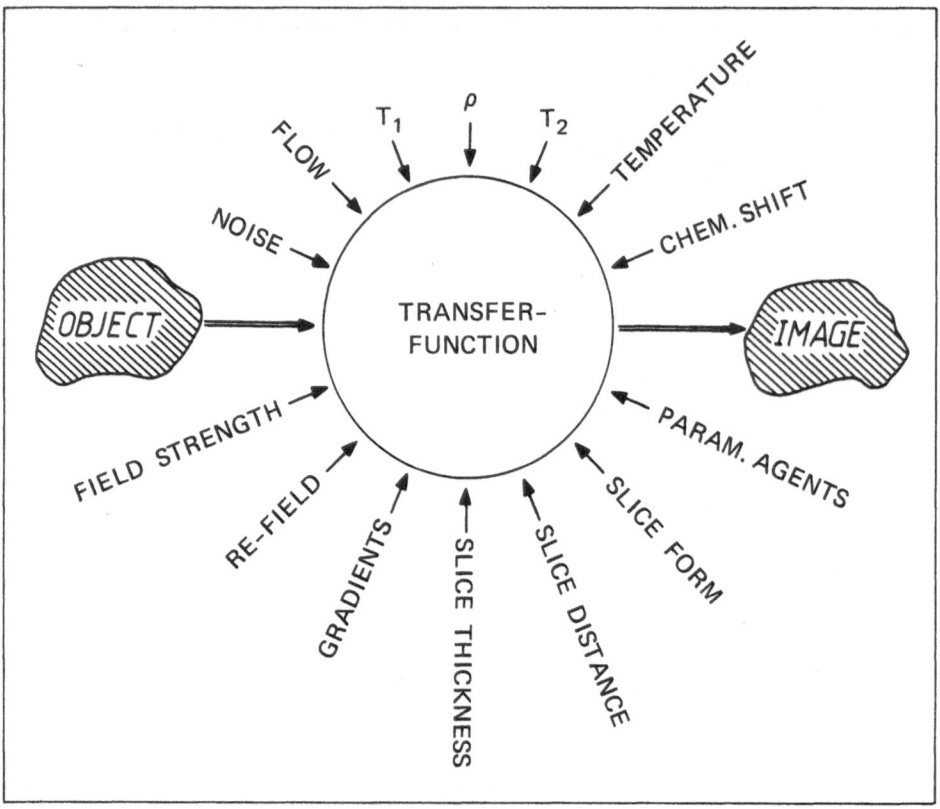

Fig. 10.1. Intrinsic and extrinsic parameters in NMR.

the examination do have a considerable influence on imaging, but they are not independent parameters because they depend on both RF field and RF field gradients.

Several of the intrinsic parameters can be influenced by extrinsic parameters. The relaxation times T_1 and T_2 as well as chemical shift imaging depend on the field strength of the magnet system. Flow phenomena are dependent on the chosen pulse sequences, i.e. RF field and gradients.

Figure 10.2 demonstrates qualitatively the improvement of the signal-to-noise ratio with an increase in field strength, whereas the depth of penetration for high frequency fields decreases.

With higher field strengths the T_1 values of tissues approach the value of water, whereas T_2 is only slightly influenced.

The image series in Figs. 10.3–10.8 exemplify the possibility of emphasising single intrinsic parameters in the NMR tomogram. Figure 10.3 shows a T_2 weighted, Fig. 10.4 a T_1 weighted and Fig. 10.5 a proton density weighted NMR image. A certain area in Fig. 10.6 lacks NMR signals owing to blood flow phenomena. Finally. Fig. 10.7

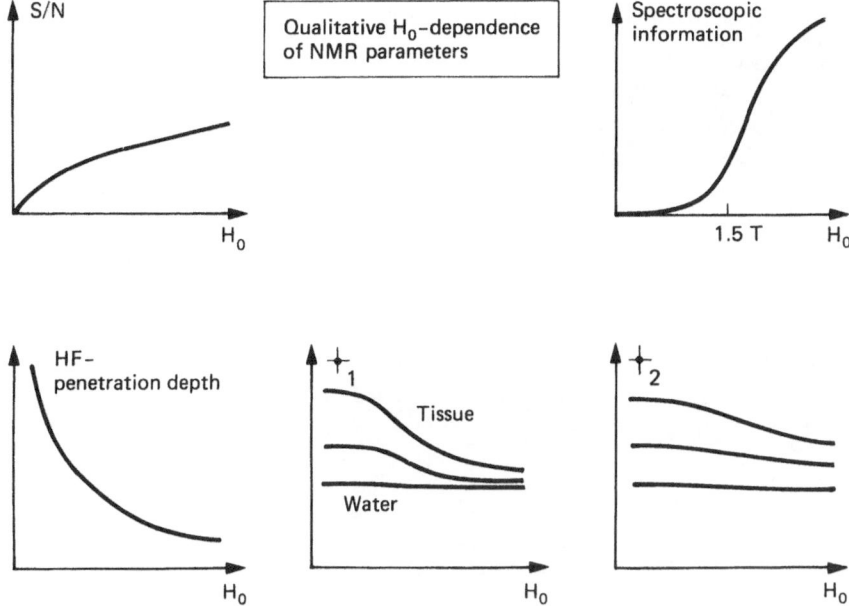

Fig. 10.2. Qualitative course of NMR parameters with field strength.

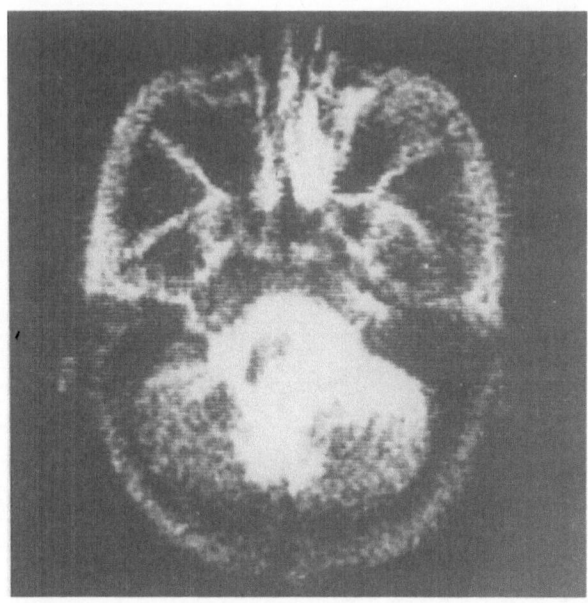

Fig. 10.3. T_2 weighted NMR tomogram.

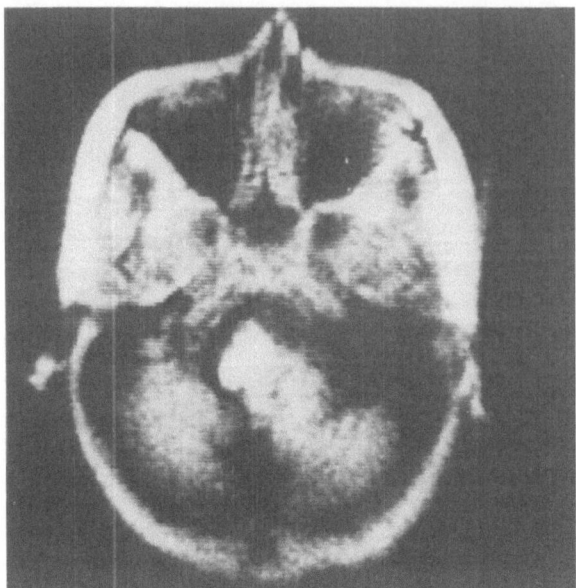

Fig. 10.4. T_1 weighted NMR tomogram.

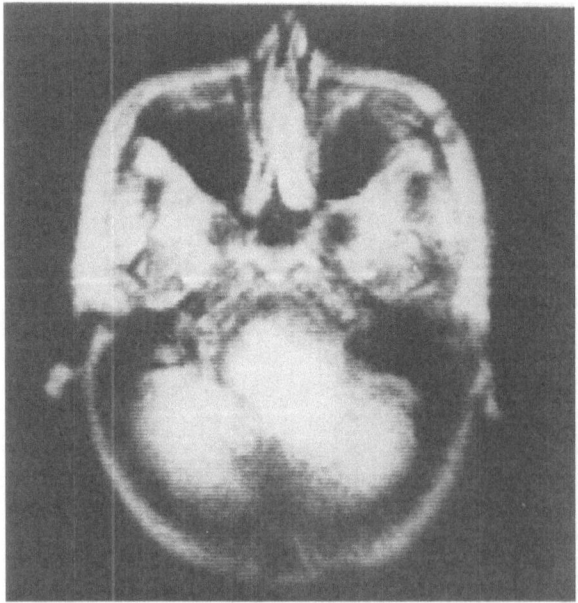

Fig. 10.5. A proton density weighted NMR tomogram.

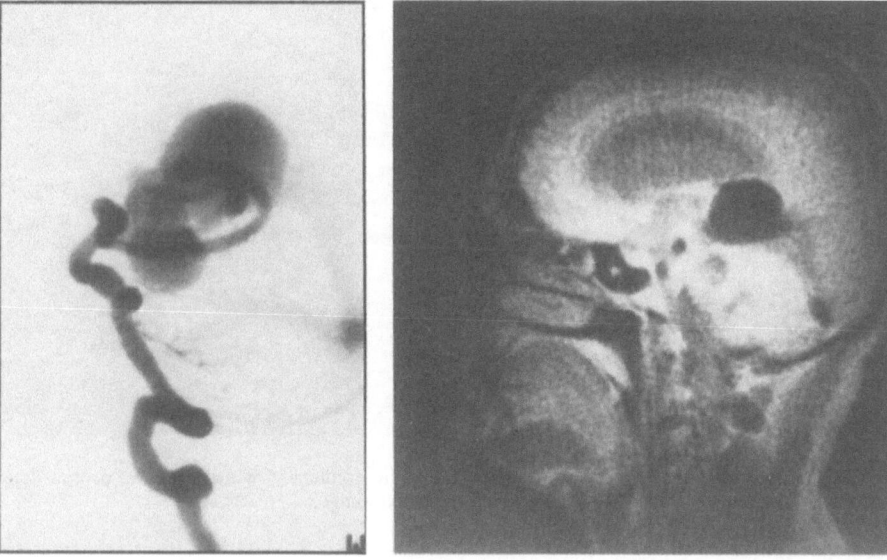

Fig. 10.6. **a** An intra-arterial DSA and **b** a parasagittal NMR tomogram of an angioma located in the posterior cranial fossa. The areas of low signal intensity can be explained by flow phenomena in the vascularised part of the arteriovenous malformation.

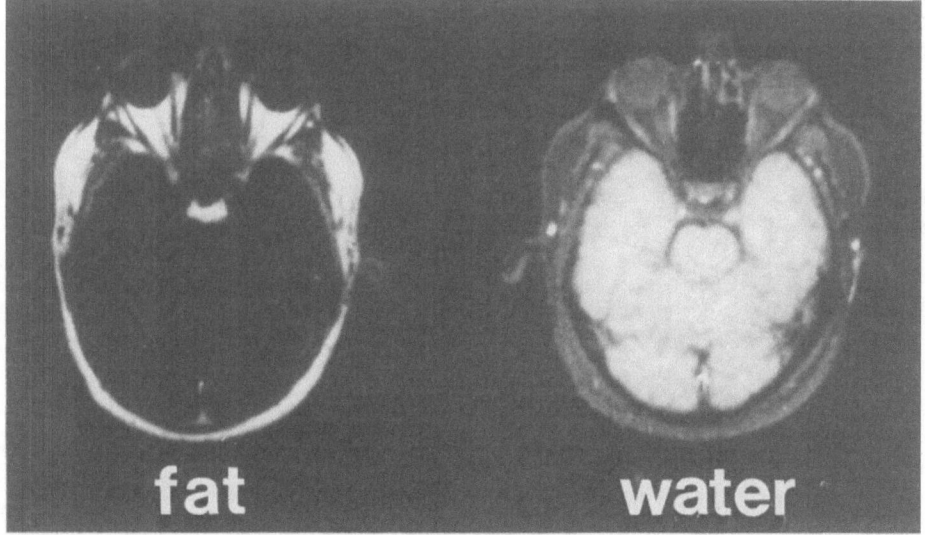

Fig. 10.7. Illustration of the discrimination between fat and water by means of chemical shift imaging.

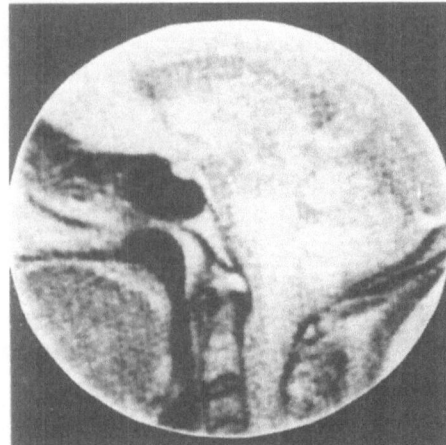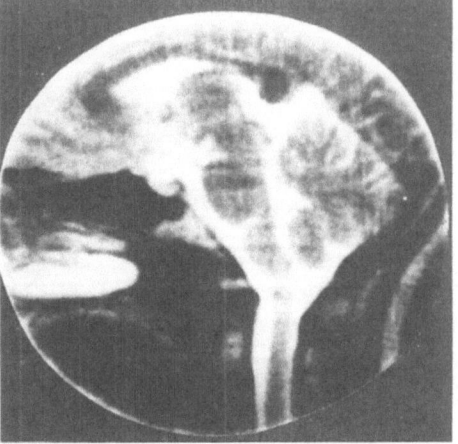

Fig. 10.8. Zoom effects by means of changing the slope gradient. On the left side: proton density weighted NMR image. On the right side: T_2 weighted NMR image.

demonstrates the effects of chemical shift imaging due to high field strengths. Figure 10.8 presents zoom effects achieved by changing the degree of gradient slope. All images except the one showing chemical shift imaging were generated with a resistive 0.14 Tesla magnet system.

In contrast to the generally fixed field strength of NMR magnet systems, the pulse sequences and gradients generated by means of the RF field can be changed by the examiner. The pulse sequences differ mainly both in the initial phase angle (α) of the magnetisation vector, usually obtained by different pulse length of the RF field, and the time of signal read out, for example by means of a 90° pulse.

In Fig. 10.9 the duration of high frequency pulses is chosen so that the magnetisation of the protons is rotated by a 90° pulse out of the main magnetic field direction. This is a saturation recovery sequence where the energy levels are equally occupied by the 90° pulse, with an example shown in Fig. 10.10. Due to the difference

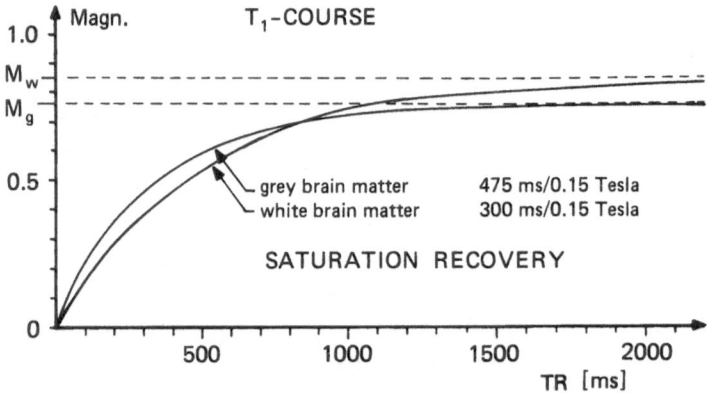

Fig. 10.9. Course of the T_1 values for grey and white brain matter at a defined field strength depending on the repetition time TR and the longitudinal relaxation time with saturation recovery technique.

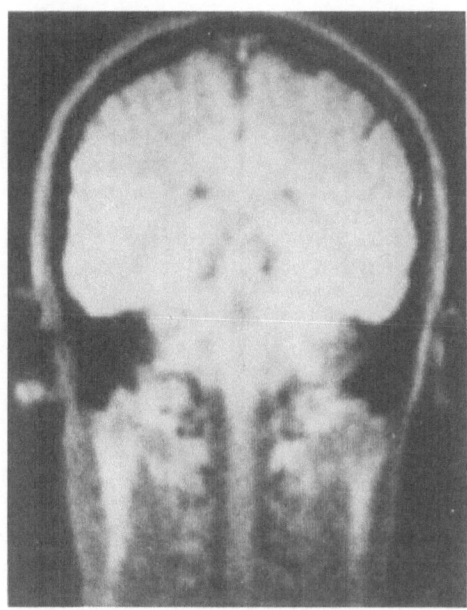

Fig. 10.10. NMR tomogram corresponding to Fig. 10.9 with the pulse sequence parameters TR = 1260 ms and TE = 49 ms.

in the proton density of the chosen pulse sequence parameters (TR = 1260 ms, TE = 49 ms) the grey matter of the brain appears dark, whereas the white matter appears bright.

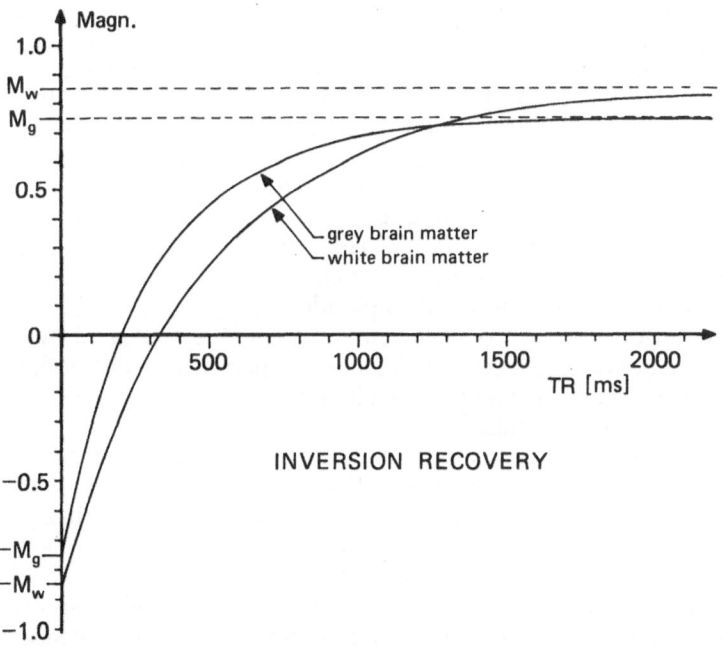

Fig. 10.11. The T_1 course (similar to Fig. 10.9) with inversion recovery technique.

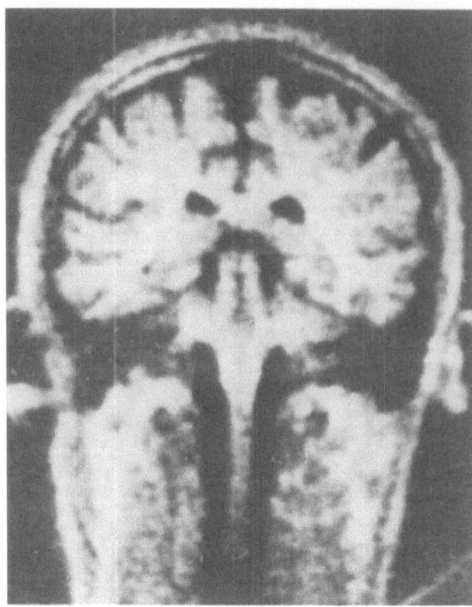

Fig. 10.12. An NMR tomogram in inversion recovery technique with the pulse sequence parameters TR = 1600 ms, TI = 400 ms, and TE = 49 ms.

Figure 10.11 illustrates the inversion recovery technique with a phase angle of 180°. Using those pulse sequences the NMR images are weighted by T_1 which represents the relaxation time the "excited" protons need to return to their initial behaviour in the magnetic field. There is a clear differentiation between grey and white brain matter, with inverted intensity in the corresponding NMR tomogram (Fig. 10.12). Despite small differences in the T_1 relaxation times, contrast improves due to the signal intensities turning from negative to positive values. The contrast is maximal if both tissues are equal in signal intensity with different sign.

A further technology utilises spin echo pulse sequences generated by means of the Carr-Purcell-Meiboom-Gill sequence (CPMG) (Fig. 10.13). After a saturation recovery or an inversion recovery sequence, several successive 180° stimulated pulses are produced with varying TE. The maximal amplitude of the echoes — caused by refocussing — is found on a "real" T_2 decay curve which is not falsified by the remaining inhomogeneities of the magnetic field. These experiments allow both the perception of the relaxation time T_2 and the pursuit of the T_2 decay over a period of up to 350 ms or longer after the 90° pulse.

Figure 10.14 shows four images of this echo train with 24 echoes, where six sequential echoes were added to form one image. The different shades of grey in the first image indicate that ρ and T_1 are emphasised, whereas short, middle and long T_2 values are represented in the remaining three images. It is in the T_2 weighted tomograms that imaging contrast is particularly distinctive.

Figure 10.15 shows schematically the problems which arise by using spin echo sequences. Totally different NMR parameters can lead to the same signal intensities and contrasts.

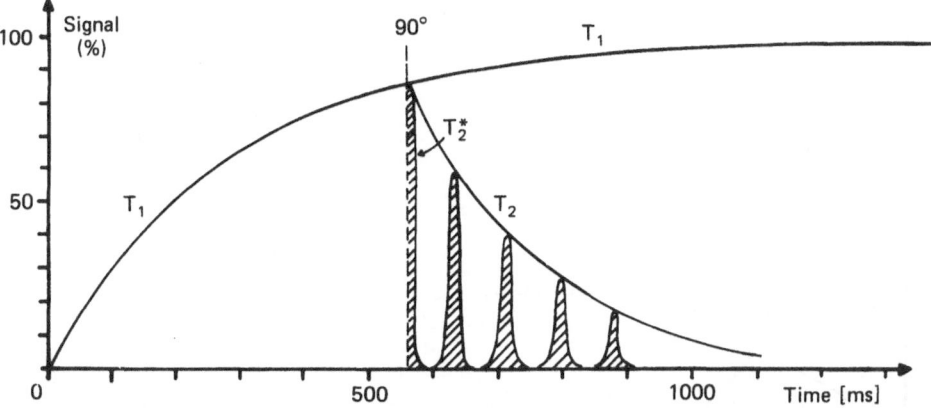

Fig. 10.13. Schematic representation of the Carr-Purcell spin echo sequence modified by Meiboom-Gill (CPMG). Twenty-four echoes are combined to form four groups (images) of six echoes each. T_2^* takes into account the magnet field inhomogeneities; the envelope of the echoes decays with the time constant T_2 influenced by the spin-spin interaction. Single echoes can be considered as sample points for the assessment of the T_2 decay.

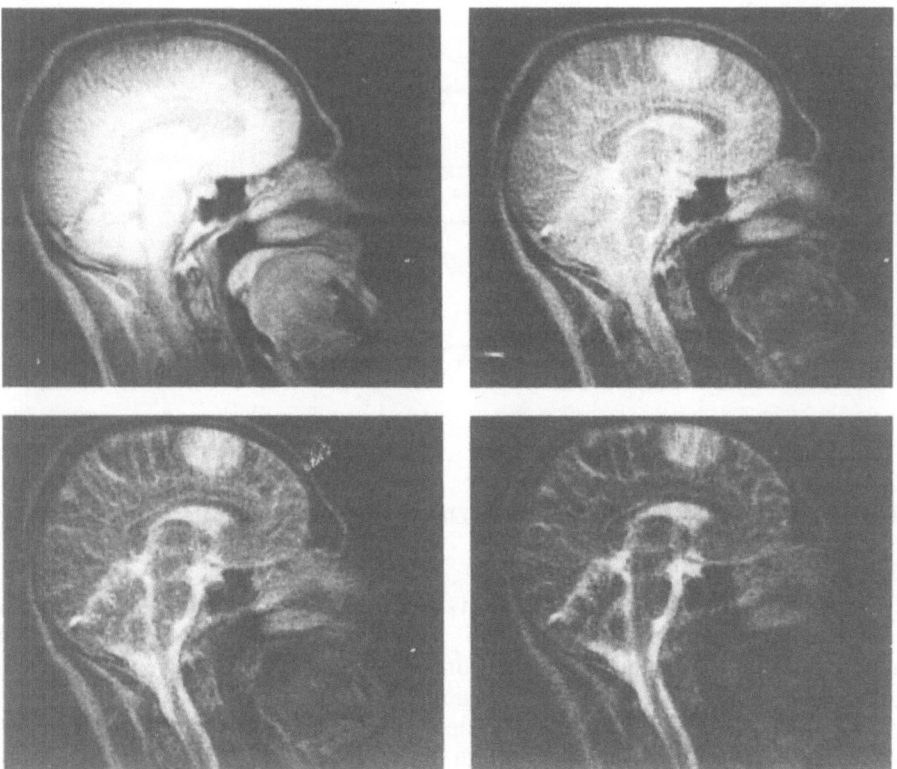

Fig. 10.14. Sequence of images generated by CPMG technique with different emphasis on ρ, T_1 and T_2. A standardised TR of 1260 ms with a τ of 7 ms was chosen, resulting in four TE times of 49, 133, 217 and 301 ms.

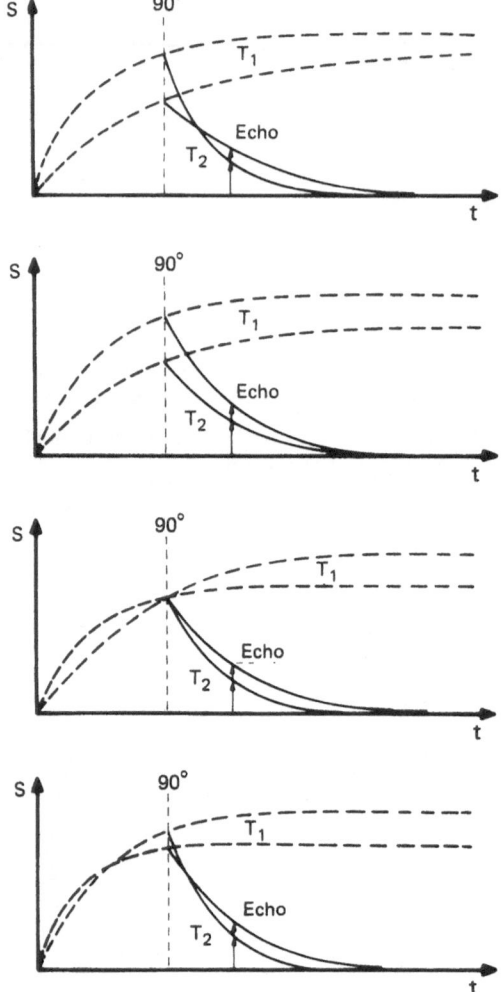

Fig. 10.15 Possible intensity course of two pixels in an NMR image. Echo signals of equal intensity can be produced by means of totally different parameter combinations.

Quantitative Evaluation and Image Synthesis

Repeated measurements with different repetition times allow a quantitative measurement of the parameters T_1 and T_2, as illustrated in Fig. 10.16 for different structures of brain tissue: corpus callosum, thalamus, pons, concha, lingua and cerebrospinal fluid (CSF). The single "regions" are shown in Fig. 10.17.

At the points of intersection of the T_2 decay and T_1 recovery curves, contrast can be lost. This phenomenon can lead to non-discrimination of a tumour from normal brain tissue in spite of good imaging quality. Several images are needed to determine whether dark areas in the image are an expression of low proton density or long T_1, short T_2 or a combination of these effects.

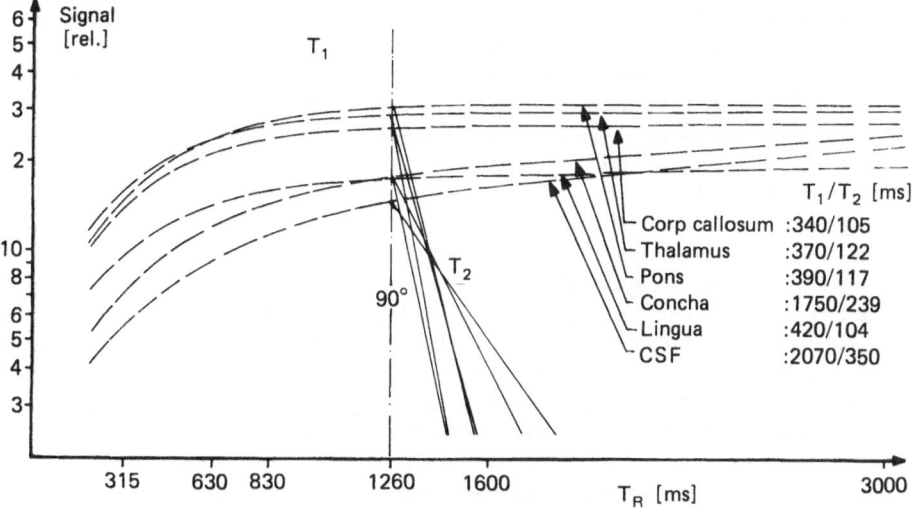

Fig. 10.16. Measured T_1 and T_2 course of normal tissue of the brain with the measured parameter values.

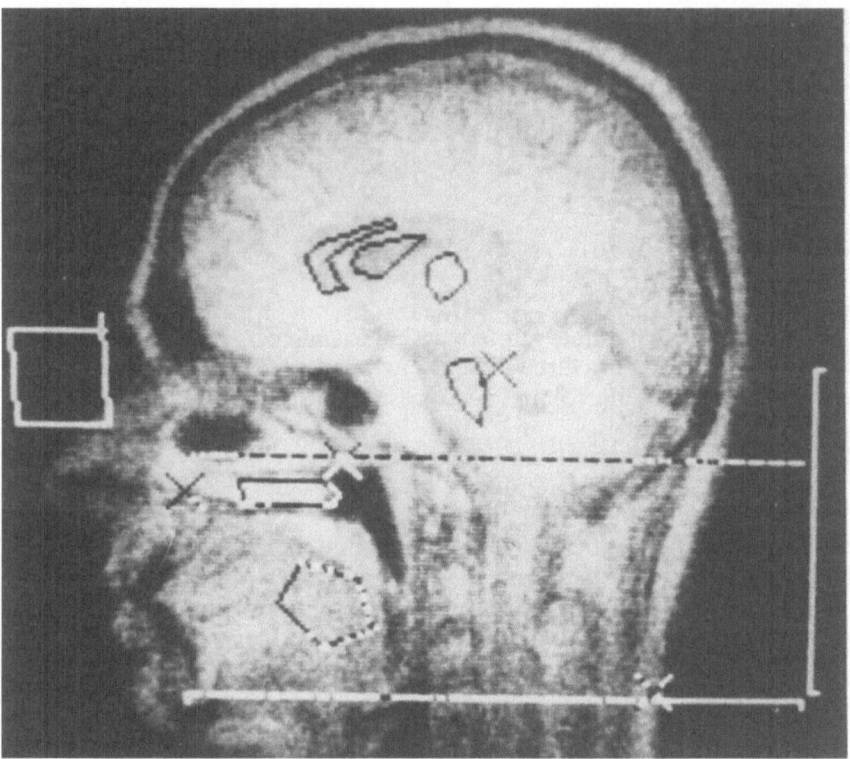

Fig. 10.17. Parts of the brain (corpus callosum, thalamus, pons, concha, lingua and cerebrospinal fluid) used for the T_1 and T_2 parameter measurements of Fig. 10.16.

Since extensive parameter variations during a patient's examination are impossible due to long measuring times, T_1, T_2 and the distribution of proton density in a cross-sectional plane are determined by means of two CPMG sequences with different repetition times. Subsequently these data provide the possibility of synthesising further images (Bielke et al. 1984) in the computer with pulse sequence parameters which could not be measured.

Figure 10.18 serves as an example of a patient suffering from a glioblastoma. The lower image series represents the single images for the B series acquired by the eight echoes of the CPMG sequence with the saturation recovery technique.

If an exponential decay of the T_2 signals is assumed, pure T_2 images can be calculated from the two repetition times A and B. By means of extrapolating the T_2 decay curves to the point of the T_1 curve where the 90° stimulation pulse occurs, NMR images are obtained providing information about the T_1 relaxation time and proton density ρ without a contribution of the T_2 relaxation times. With the help of these two images a pure T_1 as well as a pure proton density image can be calculated — a separation which would not be achievable with measuring techniques only.

The perifocal oedema surrounding the tumour is clearly visible in the pure T_1 image. The T_2 image shows the tumour as a ring, whereas the proton density image mainly delineates the central necrosis of the tumour.

Carr-Purcell spin echo sequences are extremely useful in making the best use of the specific information contained in the sequential NMR tomograms (Higer et al., to be published; Meves et al. 1984). We believe that because of the electronic increase in contrast, the application of paramagnetic substances as contrast media could be restricted to a small number of patients.

Tissue Characterisation

Finally, two examples will illustrate our biophysical examinations. The representation of proton density, T_1 and T_2 in a three-dimensional space allows a tissue classification achieved by assigning a specific colour to each tissue which can be overlaid on a black and white image (Bielke 1984).

Ideally, tissues are differentiated by characteristic colours, i.e. red is for haemorrhage, green for oedema, turquoise for white matter, lilac for grey matter, brown for CSF and yellow for fat (Fig. 10.19).

In order to carry out a classification of tissues in different groups, it is also possible to use the spatial density distributions of the T_2 parameters.

Although at present T_2 analysis alone does not allow discrimination between pathological changes such as inflammation, degeneration and tumour, the quantifiable modification of these images can be of great significance for individual follow-up, i.e. after percutaneous irradiation.

The use of quantitative procedures for tissue characterisation requires a high measurement accuracy of the NMR parameters.

Figure 10.20 shows the measurement accuracy of our NMR tomograph vis-à-vis T_1 and T_2 measured with a special phantom of the German Society of Medical Physics.

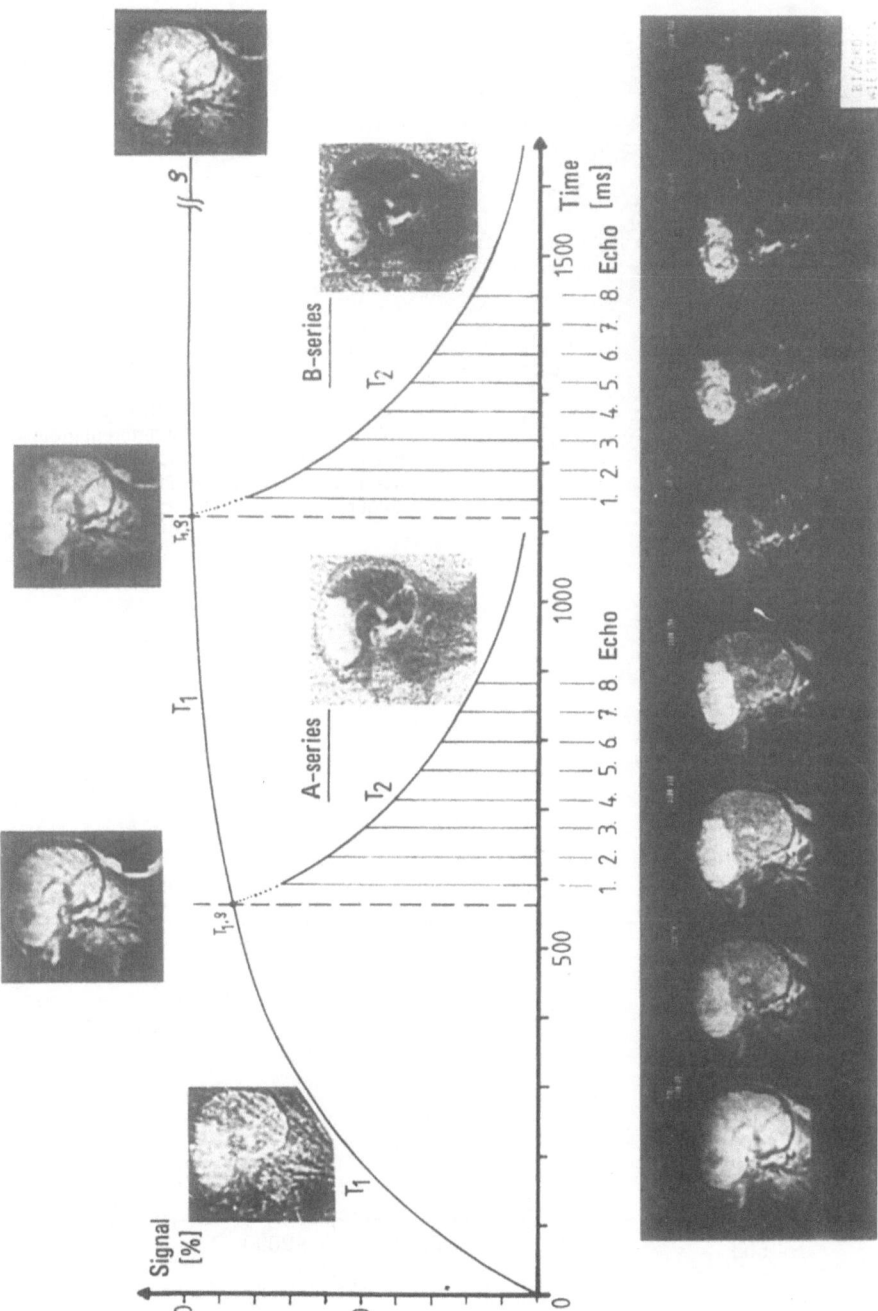

Fig. 10.18.   Procedure for the calculation of pure parameter images (T_1, T_2, ρ).

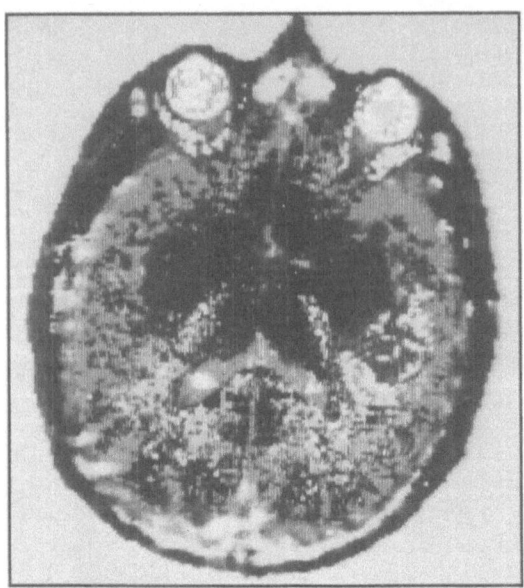

Fig. 10.19. Classification by means of coloured representation of different groups in a three-dimensional space. Each tissue is automatically characterised by a particular colour.

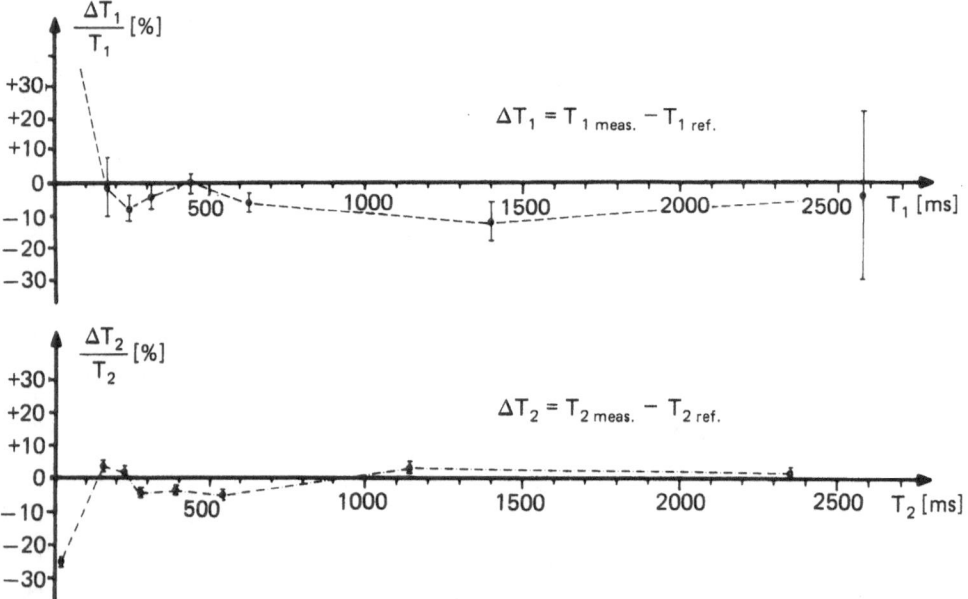

Fig. 10.20. Measurement accuracy of T_1 and T_2 of the NMR system in use at the authors' institution. The measurements were done with a calibrated phantom of the German Society of Medical Physics (DGMP).

Future Considerations

The ultimate objective of future research work and further clinical evaluation should be the optimisation of pulse sequences in NMR imaging as well as the acquisition of reliable standardised parameters for solving various diagnostic problems (Pfannenstiel and Meves 1984). Unlike the physician working in the field of X-ray computed tomography, it is of great importance that the person performing NMR examinations understands the influence of NMR parameters on imaging in order to take utmost advantage of the possibilities offered by the fascinating technology of nuclear magnetic resonance.

Acknowledgement. This work was supported by the Federal Ministry of Research and Technology.

References

Bielke G (1984) Einfluß der NMR-Parameter auf die Bildgebung und Ansätze zu deren quantitativen Darstellung. In: Schmidt T (ed) Medizinische Physik. Hutchig, Heidelberg, pp 471–478

Bielke G, Meves M, Miendl S, Brückner A, Rinck PA, von Seelen W, Pfannenstiel P (1984) A systematic approach to optimisation of pulse sequences in NMR-imaging by computer simulations. In: Esser PD, Johnston RJ (eds) Technology of nuclear magnetic resonance. Society of Nuclear Medicine, New York, pp 104–117

Higer HP, Meves M, Pfannenstiel P (to be published) Kontrastverstärkung durch T_2-betonte Multiecho-Pulsefrequenzen. Zweites Münchener NMR-Symposium, Oktober 1984. Nuklearmedizin

Meves M, Bielke G, Rinck PA, Brederhoff J, Bieler UE, Pfannenstiel P (1984) Modifizierte Spin-Echo-Sequenz in der NMR-Tomographie. In: Adam WE, Schmidt HAE (eds) Verhandlungsbericht der 21. Internationalen Jahrestagung der Gesellschaft für Nuklearmedizin Europa. Schattauer, Stuttgart, pp 68–71

Pfannenstiel P, Meves M (eds) (1984) Die NMR-Tomographie – Klinischer Einsatz und Wirtschaftlichkeit. Georg Thieme Verlag, Stuttgart

11. A Correlative Approach to Nuclear Medicine and NMR Imaging

M.P. Sandler, M.V. Kulkarni, J.A. Patton, C. Leon Partain
and A. Everette James

Introduction

Although magnetic resonance imaging is in the early stages of development, the ability to provide anatomical information comparable to both X-ray computed tomography and ultrasound, as well as to provide functional imaging information comparable to nuclear medicine examinations has been demonstrated (Margulis et al. 1983). The information contained in an NMR image is determined by four biophysical parameters related to the hydrogen nucleus: proton density relaxation times; spin-lattice relaxation time (T_1); spin-spin relaxation time (T_2); and motion due to flow. At this stage of NMR technology, clinical studies utilise the hydrogen atom (proton) for imaging.

Certain disadvantages are associated with NMR imaging. In the first instance, ferromagnetic material contained in certain life-sustaining apparatus, surgical clips, prostheses and pacemakers may be absolute contra-indications to the performance of NMR examinations. In addition, the long scanning times required may result in suboptimal images due to motion artefact. Prolonged imaging times at present limit NMR's ability to perform high temporal resolution dynamic imaging as compared to nuclear medicine. The magnetic field produced by the NMR scanner may interfere with nearby electronic apparatus. Similarly, the presence of material containing ferrous or ferromagnetic particles might affect the magnetic field and influence the site selection and installation of NMR scanners. Unlike nuclear medicine, NMR cannot be performed as a portable study.

Nuclear magnetic resonance imaging has proved to be a valuable modality in studying intracranial, cervical, thoracic, abdominal and skeletal diseases. In addition to providing good spatial resolution, NMR has demonstrated excellent soft tissue contrast and early detection of pathological abnormalities. Nuclear medicine studies also provide early detection of pathological lesions, although the spatial resolution is inferior to NMR imaging. Our initial experience with a 0.5 Tesla superconducting magnet (Teslacon Technicare) indicates excellent NMR imaging potential, enhanced by the ability of varying pulse sequences to provide T_1 and T_2 weighted images which improve tissue contrast and image detectability.

Central Nervous System

Nuclear magnetic resonance has evolved as a primary imaging modality depicting many intracranial and spinal cord abnormalities. The availability of direct coronal and sagittal imaging has improved the understanding of intracranial anatomy and the localisation of intracranial pathology (Bydder et al. 1982). The brainstem and middle intracranial fossa are particularly well imaged by NMR. The diagnosis of syringomyelia in the cervical cord can be made on a sagittal NMR performed through the cervical cord in that it shows an area of decreased intensity in the middle of the cervical spinal cord and has the same signal intensity as CSF surrounding the spinal cord.

Unlike radionuclide brain scans where the isotope localisation in the intracranial lesions depends on the integrity of blood–brain barrier, NMR imaging identifies abnormalities because of increase in mobile hydrogen atoms. Araki and his colleagues suggested that malignant neoplasms have longer T_1 relaxation times when compared to benign tumours (Araki et al. 1984). In our experience, it is not possible to distinguish tumour from oedema because both entities produce abnormal signals

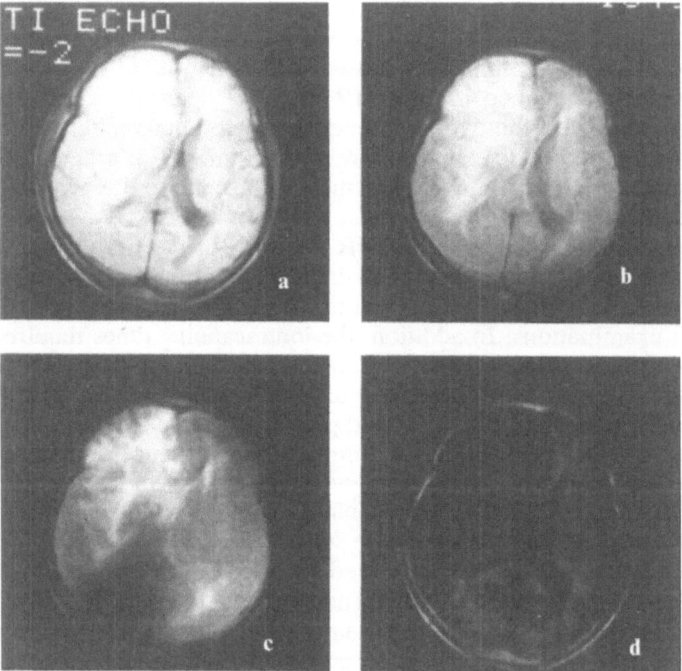

Fig. 11.1a-d. Use of different pulse sequences at the same transverse level reveals that the contrast between the frontal lobe abnormality and the normal brain is seen in SE 120/1000 (c) and IR 30/450/1500 (d) images which are T_2 and T_1 weighted images respectively. Although the SE 30/1000 (b) image does not show good contrast between the abnormality and the normal brain, the mass effect on the right lateral ventricle is better using this pulse sequence. At biopsy this abnormality was determined to be a focus of progressive multifocal leukoencephalopathy.

on T_1 and T_2 weighted images. In a patient with progressive multi-focal leukoencephalopathy, the NMR image shows an area of increased signal intensity in the right frontal region (Fig. 11.1). Figure 11.1 also demonstrates the use of different pulse sequences in detecting the lesion. With spin echo images with short TE (30 ms) and TR (1000 ms), the mass effect on the right lateral ventricles and the midline shift are well depicted (Fig. 11.1b). With increased TE, increasing signal intensity is noted in the right frontal region. Using an SE 120/1000 image (Fig. 11.1c bottom left), the abnormality is seen as an intense signal, although the lateral ventricles are not well depicted using this pulse sequence. These T_2 weighted images (prolonged TE) have been found most sensitive in detecting abnormalities. Figure 11.1d represents inversion recovery pulse sequence. Radionuclide brain scan can also demonstrate the abnormality, but the mass effect and midline shift are better demonstrated with NMR. Inflammatory lesions may be detected on NMR images, and preliminary work with paramagnetic contrast agents in animal brain abscess models using gadolinium-DTPA contrast enhanced MRI studies at Vanderbilt have indicated that these studies are superior to nuclear medicine (Runge et al. 1984). The use of intravenous paramagnetic contrast agents may help us to make the differentiation between neoplastic and inflammatory lesions.

Neck

Thyroid

Solitary Thyroid Nodules

In vitro studies on thyroid biopsy specimens performed by de Certaines and colleagues revealed abnormal T_1 and T_2 values in both benign and malignant lesions (de Certaines et al. 1982). Nodules showing increased radionuclide uptake demonstrated a marked degree of variability in T_1 values but all ($n = 10$), with the exception of a single patient, had significantly increased T_2 values compared to normal extranodular tissue. Solitary benign cold nodules ($n = 9$) all showed increased T_1 and T_2 values. T_1 and T_2 values showed considerable variation in four patients with thyroid carcinoma. An increase in T_1 was observed in two patients and a decrease in one. T_2 values increased in two cases and declined in two. There was no significant difference in the relaxation times. In a recent limited patient series the T_1 values of malignant and benign thyroid disease could not be reliably distinguished (unpublished data). Figures 11.2 and 11.3 illustrate the NMR images of a benign thyroid adenoma and thyroid carcinoma residing in a thyroglossal duct cyst.

Colloid cysts examined in vivo also exhibit prolonged T_1 values characteristic of simple fluids. Haemorrhage into a cyst should lower the T_1 value; this hypothesis, however, remains to be proven. Although generally prolonged, adenomas exhibited a wide range of T_1 values, which encompassed that of the thyroid carcinomas examined. Care must be exercised in the use of T_1 because calibration of an NMR imager must be continually performed to assure precise determination of T_1. In addition, the absolute value of T_1 is dependent upon the strength of the magnetic field utilised. Thus, comparison of values between instruments with different magnetic field strengths and resonant frequencies is complex.

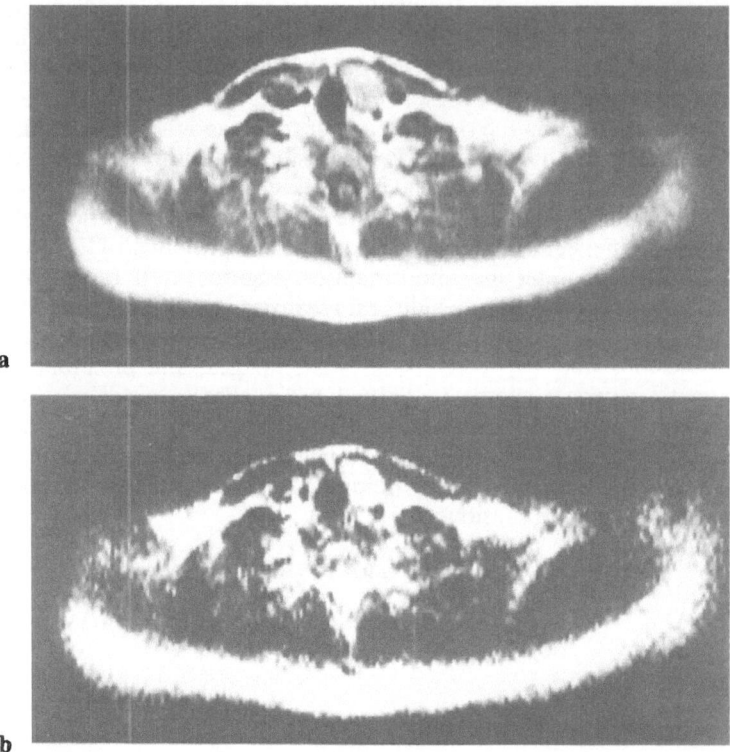

a

b

Fig. 11.2. **a** Transverse section through a benign hypertrophic nodule in the left lobe of the thyroid with spin echo 30/500 pulse sequence. **b** Same transverse section with spin echo 120/2000 pulse sequence.

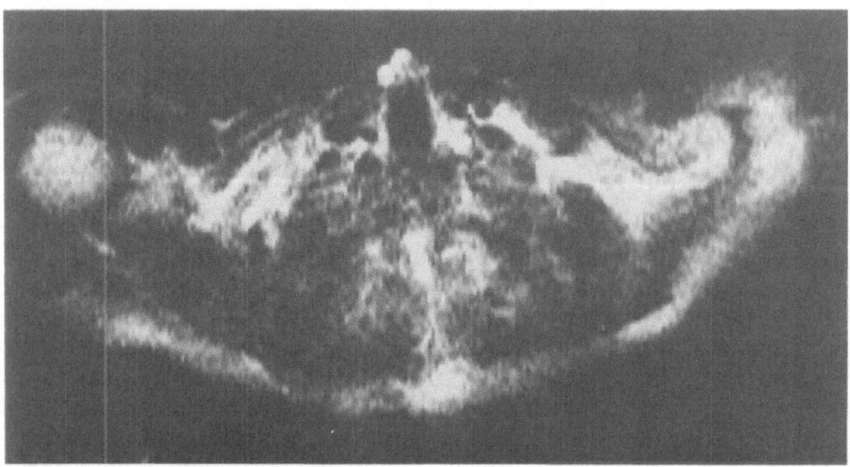

Fig. 11.3. A thyroglossal duct cyst with spin echo 120/2000. At surgery this patient was noted to have a papillary carcinoma (Sandler et al., to be published).

Retrosternal Goitre

Although substernal aberrant thyroid accounts for only 10% of all mediastinal masses, it remains one of the major diagnostic considerations in the assessment of these abnormalities. Goitres in the superior mediastinum generally arise from one or both lower poles of the thyroid gland or from the isthmus. Although clear continuity between the cervical and intrathoracic components should be present in cases of mediastinal goitre extension, the connection may only be a narrow fibrous band or vascular pedicle. In such cases, as well as in the presence of primary intrathoracic goitre, lack of continuity between the cervical gland and the thoracic mass does not exclude the diagnosis of mediastinal goitre. The preoperative diagnosis of a thyroidal mediastinal mass frequently requires the use of sophisticated imaging techniques, either alone or in combination. These include radionuclide thyroid scintigraphy, computed tomography (CT) and possibly nuclear magnetic resonance (NMR) imaging.

Radionuclide scanning has been the standard method for evaluating whether or not a mediastinal mass represents functioning thyroid tissue. False-negative thyroid scans in mediastinal goitre may occur when there is too little uptake of radioactive iodine by the goitre either due to low iodine concentrating capacity of the tissue or recent exposure of the patient to exogenous iodines. Although false positive thyroid scans theoretically may occur in rare mediastinal teratoma similar to ovarian struma, only one false positive scan has been reported; this was in a patient with poorly documented papillary adenocarcinoma of the lung (Fernandez-Ulloa et al. 1976).

CT findings of intrathoracic goitre have noted the continuity with the cervical gland, focal calcifications, high non-contrast attenuation values of the goitre, and postintravenous enhancement. These features, although suggestive of intrathoracic goitre, are not specific for the diagnosis of a thyroid mediastinal mass. The early enhancement of intrathoracic masses following intravenous contrast more likely represents the vascularity of the mass rather than significant trapping and organification by the organ. This is supported by cases reported in the literature demonstrating marked contrast enhancement in mediastinal masses other than thyroid tissue. Contrast-enhanced CT as the initial diagnostic procedure may occasionally prove counterproductive in the non-invasive workup of the mediastinal mass because expansion of the body pool of iodine by contrast infusion will interfere with normal thyroid uptake beyond the blood pool phase, a scintigraphic finding specific for mediastinal thyroid. Nuclear magnetic resonance (NMR) imaging is a technique which can produce high-resolution tomographic or three-dimensional images of the mediastinum (Runge et al. 1984; de Certaines et al. 1982; Sandler et al. 1984). The success of NMR imaging in the mediastinum is due to its superior ability to differentiate vascular structures from solid hilar or mediastinal masses without the use of contrast agents (Figs. 11.4, 11.5).

Graves' Disease

Graves' disease can be distinguished on the basis of T_1 values from normal thyroid tissue. The physiological basis of the prolonged T_1 values of Graves' disease remains unexplained.

ANTERIOR

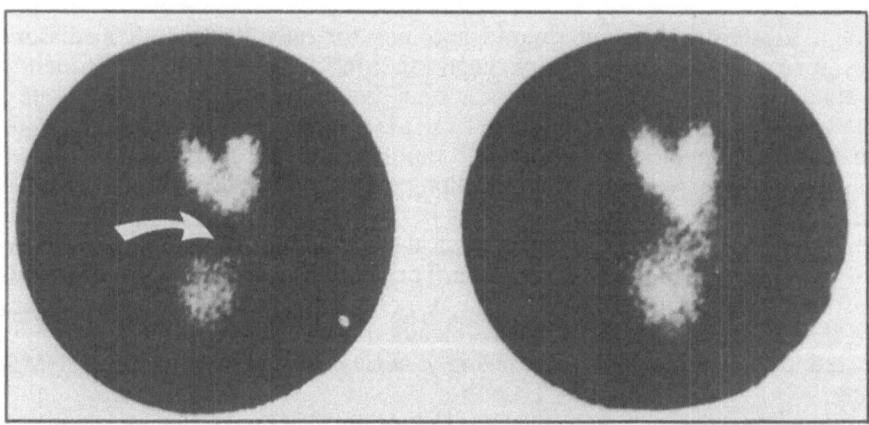

1-123

Fig. 11.4. I-123 scintigram demonstrates asymmetrical enlargement of the thyroid gland, with evidence of retrosternal extension into right mediastinum. Suprasternal notch (*arrow*) has been marked with lead absorber (Sandler et al. 1984).

Parathyroid

Nuclear Medicine: Technetium 99m/Thallium 201 Subtraction

Ferlin et al. and Young et al. have recently described the successful utilisation of combined Tc-99m and Tl-201 subtraction imaging for the localisation of parathyroid adenoma in patients with primary hyperparathyroidism (Ferlin et al. 1983; Young et al. 1983). Modifications of this technique have been described by Basarab et al. (1985).

There are several advantages of Tc-99m/Tl-201 subtraction for the detection of parathyroid adenoma. It is a non-invasive technique and has a high sensitivity (95%) and specificity (94%) for the detection of parathyroid adenomas. In addition, it has the potential to identify both aberrant and ectopic glands (Fig. 11.6). The disadvantages of this technique include its inability to determine adenoma depth and its inability to detect four gland hyperplasia with any degree of accuracy. False positive results have been reported in patients with thyroid nodules including hypertrophic and malignant nodules, multinodular goitre and sarcoid lymph nodes.

Modifications to improve the sensitivity include taking oblique views, administration of oral phosphates (1 g/day in divided doses for 3 weeks) and I-123/Tl-201 subtraction. A disadvantage of I-123/Tl-201 subtraction is that salivary glands will routinely appear positive since iodine does not accumulate significantly in the salivary gland, while thallium does.

Nuclear Magnetic Resonance Imaging

Nuclear magnetic resonance (NMR) imaging is suitable for demonstrating neck pathology since it offers excellent soft tissue contrast without the need for

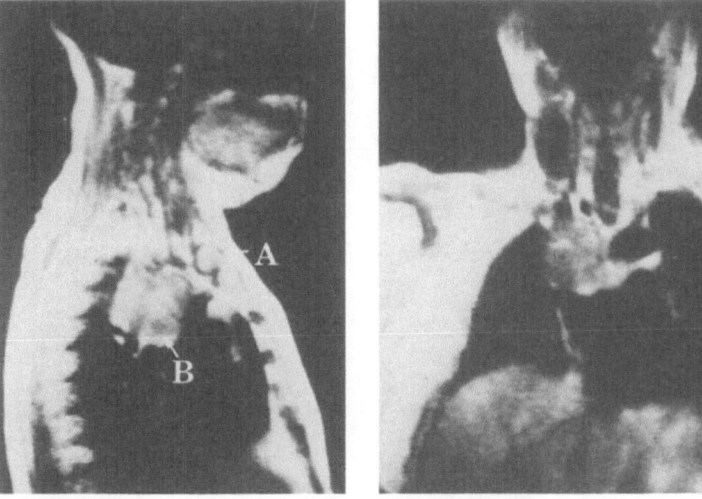

SAGITTAL   CORONAL

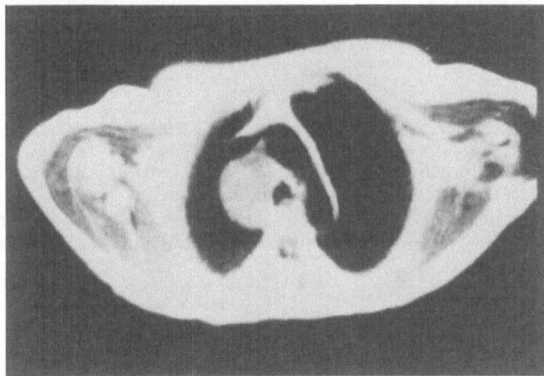

TRANSVERSE

Fig. 11.5. NMR image using spin echo technique with TE=30 ms and TR=1.0 ms. *A*, inferior pole of right thyroid lobe; *B*, midline mediastinal paratracheal mass (Sandler et al. 1984).

intravenous contrast material. It is resistant to motion related artefacts and can be used to examine the entire neck and thoracic inlet. Although the spatial resolution of NMR is currently a limiting factor, the use of specialised surface coils provides imaging with sharp detail in a superficial structure such as the parathyroid glands (Fig. 11.7). A variety of parathyroid lesions have been identified using NMR, all of which were greater than 1 cm in size. While the lesions could be separated from adjacent structures, it was not possible to distinguish reliably between the various types of parathyroid tumours. NMR has the potential to become the most accurate imaging modality in the investigation of patients with parathyroid disorders. However, its role in the routine preoperative evaluation of patients with suspected hyperparathyroidism remains to be established.

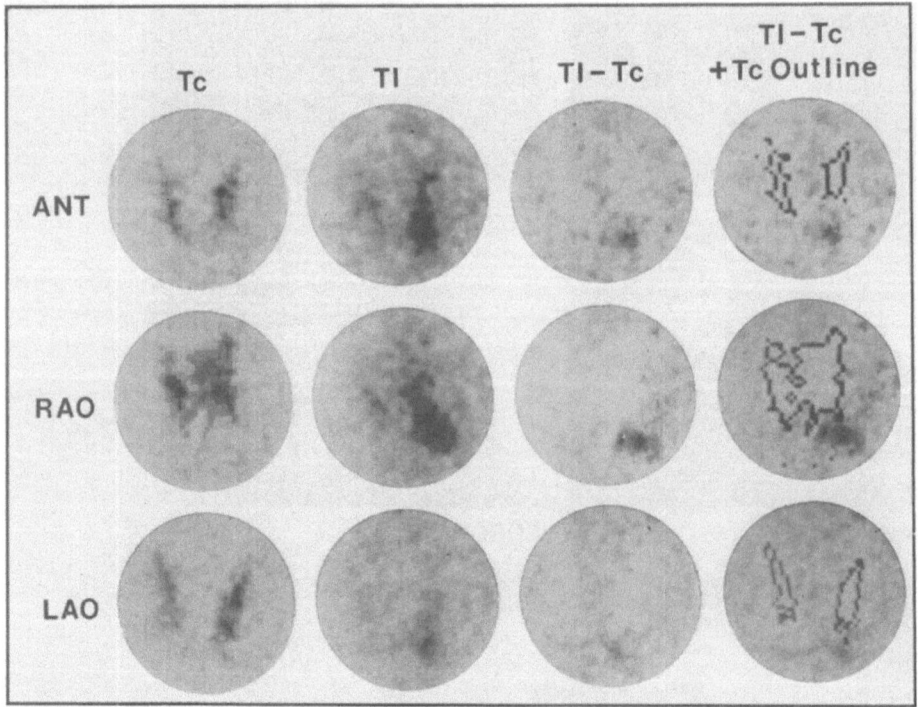

Fig. 11.6. Tc-99m, Tl-201 and Tc-99m/Tl-201 subtraction images in the anterior (*ANT*), left anterior oblique (*LAO*) and right anterior oblique (*RAO*) projections using the pinhole collimator from a patient with a parathyroid adenoma situated at the base of the left lobe of the thyroid. Subtraction images with an overlay of the thyroid obtained from the Tc-99m image are also shown for localisation purposes (Sandler et al., to be published).

Thorax

Cardiac

Nuclear Medicine

Cardiac radionuclide imaging encompasses a variety of techniques designed to provide physiological information of value in detecting the presence and extent of cardiac disease. All entail the injection of a radiopharmaceutical, its distribution to the myocardium or throughout the blood pool and the external detection of radioactivity by means of a gamma scintillation camera. Cardiovascular parameters currently evaluated with radionuclide angiography include left-to-right shunting, right-to-left shunting, cardiac output, left and right ventricular ejection fraction, left

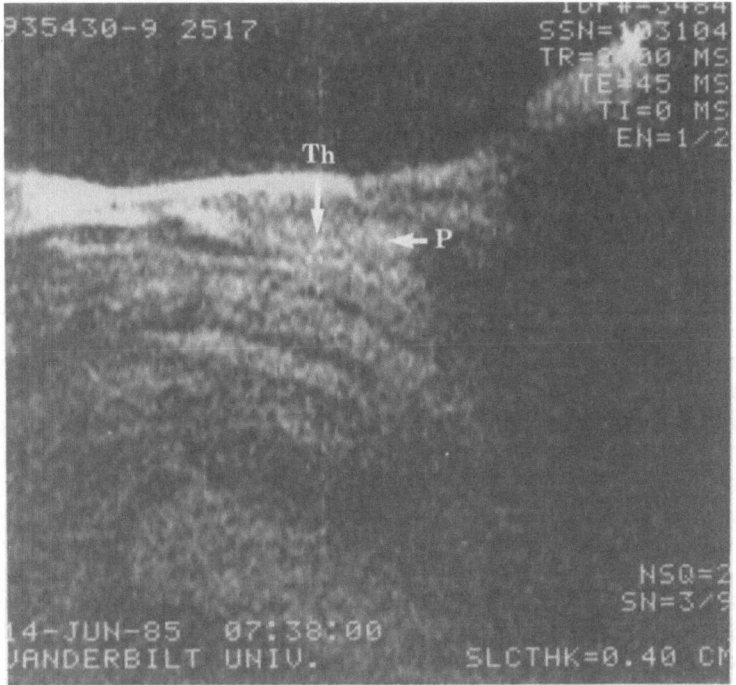

Fig. 11.7.   Left anterior oblique view of neck, MRI surface coil image, SE 45/2000, 0.4 cm slice thickness.
Parathyroid adenoma (*P*), left lobe of thyroid (*Th*) (Sandler et al., to be published).

and right ventricular end-diastolic volume, stroke volume, pulmonary blood volume
and transit time.

NMR

Imaging with nuclear magnetic resonance provides natural inherent contrast
between flowing blood and the myocardium, since nuclei of blood flowing at normal
velocities produce minimal signal relative to the surrounding cardiac tissues.
Considerable contrast also occurs between the epicardial surface and pericardial fat
since fat produces a much stronger magnetic resonance imaging signal than the
myocardium.

Gating of NMR images of the beating heart reduces, if not eliminates, motion
artefacts (Lanzer et al. 1984). Technology also now exists to obtain multiple, thinner
sections simultaneously to provide more precise anatomical detail in a shorter period
of time. Investigators have been successful in obtaining high quality gated NMR
images, demonstrating normal cardiac anatomy and cardiovascular pathology,
including a variety of congenital abnormalities (Fletcher et al. 1984a, b).

Transverse images from the aortic arch downwards to the superior aspect of the
liver provide the best initial analysis of cardiac anatomy, followed by coronal and/or
parasagittal views if considered additionally necessary. Aortic and pulmonary artery
relationships are clearly identified in transverse imaging planes, with the mid-

ascending aorta being anterior and to the right of the mid-main pulmonary artery in uncorrected D-transposition of the great vessels (D-TGVs), anterior and to the left of the pulmonary artery in corrected L-transposition of the great vessels (L-TGVs) and the two being side by side in double outlet right ventricle (DORV). Coronal images slicing from anterior to posterior clearly identify the relationship between the transposed major vessels and the ventricles.

The actual size of the main pulmonary artery and its right and left pulmonary artery branches can be analysed with exceptional clarity in the transverse plane, revealing post-stenotic dilatation in pulmonary valvar stenosis, the varying degrees of underdevelopment in severe right heart obstructive disease and patency and degree of ongoing growth after temporary surgical shunting or definitive reconstruction.

Right versus left-sided aortic arches and descending thoracic aortas are very clearly documented with transverse imaging. However, more complex presurgical analysis of aortic abnormalities (coarctation and postrepair aneurysm complication) require additional assessment of great vessel relationships that is better provided by coronal or parasagittal imaging.

Atrial septal defect (ASD) is best assessed in the transverse plane. The septum is both thin and often mobile in the region of the fossa ovalis. Similar to echocardiography, the potential for false positive ASD remains a problem. We noted a sudden focal dropout in signal intensity in the region of the ASD with prominent signals in the adjacent cranial and caudal intact septum. Repeat imaging with interweaved adjacent slices produces thinner slice evaluation (3–4 mm) and a greater accuracy of analysis. Absent cranial atrial septum followed by intact midseptum indicates sinus venosus ASD. Absent midseptum and intact uppermost and lowermost septum indicates secundum ASD.

Membranous ventricular septal defect (VSD) is well assessed with both sagittal and transverse imaging (Fig. 11.8). Multiple membranous and muscular VSDs have been imaged successfully. Differentiation between complete absence of the ventricular septum versus some residual septum remnant is readily apparent on transverse NMR. A four-chamber transverse image in the midheart defines the arterioventricular (a-v) valves leading from one or two atrial chambers to one or two ventricles. NMR generally has some difficulty in providing clear detail of most normal, and some abnormal, pulmonary and aortic valves where real-time echocardiography remains superior.

It is unlikely that NMR will replace echocardiography as the simplest and most definitive method of establishing a non-invasive diagnosis in patients with congenital cardiac malformations or will replace radionuclide angiography as the least invasive technique in the detection and quantitation of cardiac shunts. NMR is more likely to become a complementary additive non-invasive imaging procedure to answer some questions left in doubt by echocardiography (mainly extracardiac artery and vein assessments) and radionuclide angiography as a preferred follow-up imaging method in certain clinical circumstances. Angiocardiography will remain necessary to provide physiological data, i.e. chamber pressures, shunt volumes, oxygen saturations and pulmonary vascular resistance. However, NMR could negate some follow-up catheterisations in appropriate clinical circumstances. High resolution proton NMR tomography should ultimately permit the accurate evaluation of ventricular volumes, myocardial mass, and the assessment of regional wall motion and ejection fractions. Paramagnetic substances such as manganese ion may ultimately provide a basis for myocardial perfusion imaging.

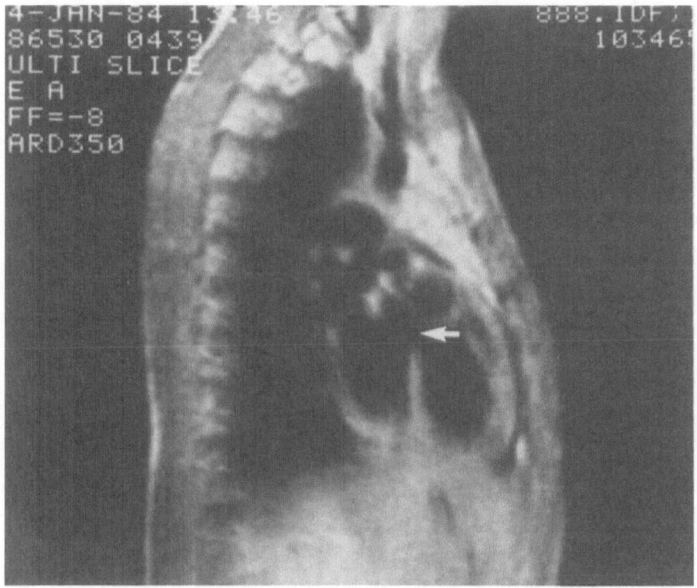

Fig. 11.8. Cardiac gated sagittal NMR image using SE 30/500 pulse sequence demonstrating membranous ventricular septal defect (*white arrow*).

Mediastinum

In our experience respiratory gating has improved image quality in some patients, although cardiac gated NMR images in the chest are more easily reproduced and adequate in almost all the patients studied with mediastinal masses. The NMR image using SE 30/500 sequence demonstrates the area of abnormal signal intensity of the right side of the mediastinum in a patient with mediastinal lymphoma (Fig. 11.9). The right atrium and the left atrium, as well as the descending aorta, can be separated from the rest of the mediastinal structures because of the decreased signal intensity within these structures, secondary to blood flow.

In a series of patients studied to evaluate mediastinal and hilar masses, NMR detected additional disease in 25% of the cases when compared with CT (Cohen et al. 1983). Magnetic resonance, however, cannot discriminate between neoplastic and inflammatory lymph node pathology by T_1 and T_2 relaxation times but can provide information regarding abnormal lymph nodes analogous to Ga-67 (gallium) citrate scans.

Abdomen

NMR of the abdomen provides complementary images to nuclear medicine and has a transverse imaging capability similar to that of single photon emission computed tomography (SPECT) and CT. Abnormalities in the region of the porta

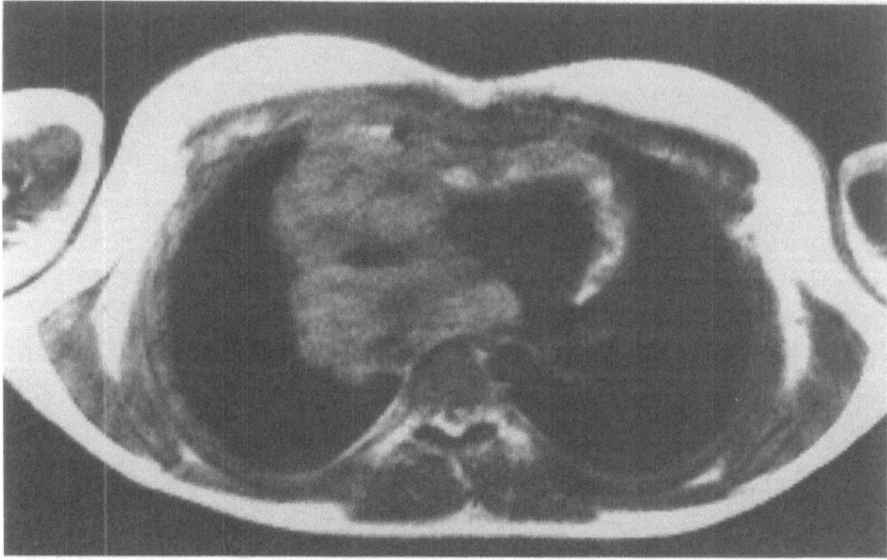

a

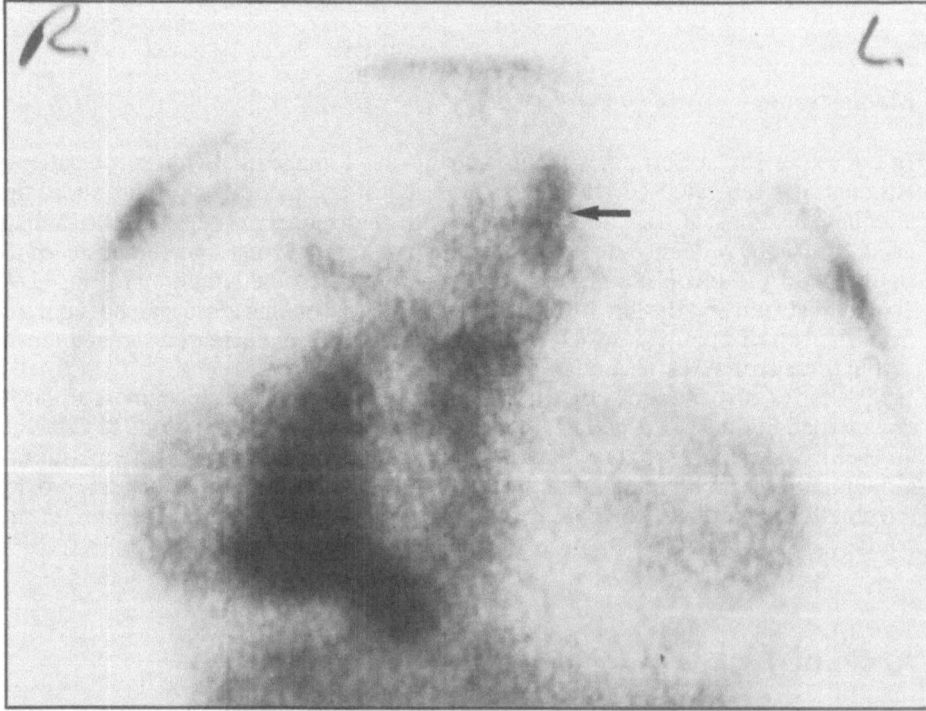

b

Fig. 11.9 **a** Cardiac gated transverse MR image using SE 30/500 sequence demonstrating abnormally increased signal in the thorax in a patient with Hodgkin's disease. **b** Anterior view of a Ga-67 citrate scan shows abnormal uptake in the right side of the mediastinum in this patient with Hodgkin's disease. Increased uptake is also seen in the left side of the neck (*arrow*).

hepatis can be better evaluated with NMR since portal vascular structures are easily identified (Doyle et al. 1982). A 33-month-old child with liver abscess was studied with NMR and an SE 30/500 image demonstrated decrease in signal intensity posteriorly in the right lobe of the liver, suggesting a lesion with increased T_1 relaxation time (Fig. 11.10a). When NMR imaging was performed at the same level using a T_2 weighted image (SE 60/1000), the abnormality was identified as an area of increased signal intensity (Fig. 11.10b). The abnormality seen on the T_2 weighted image involved a larger area of the right lobe. The increased signal intensity seen with these sequences suggested a longer T_2 relaxation time in the abscess and surrounding oedema. Ga-67 citrate imaging is also a sensitive modality in inflammatory processes but lacks specificity, and the diagnosis is usually made with

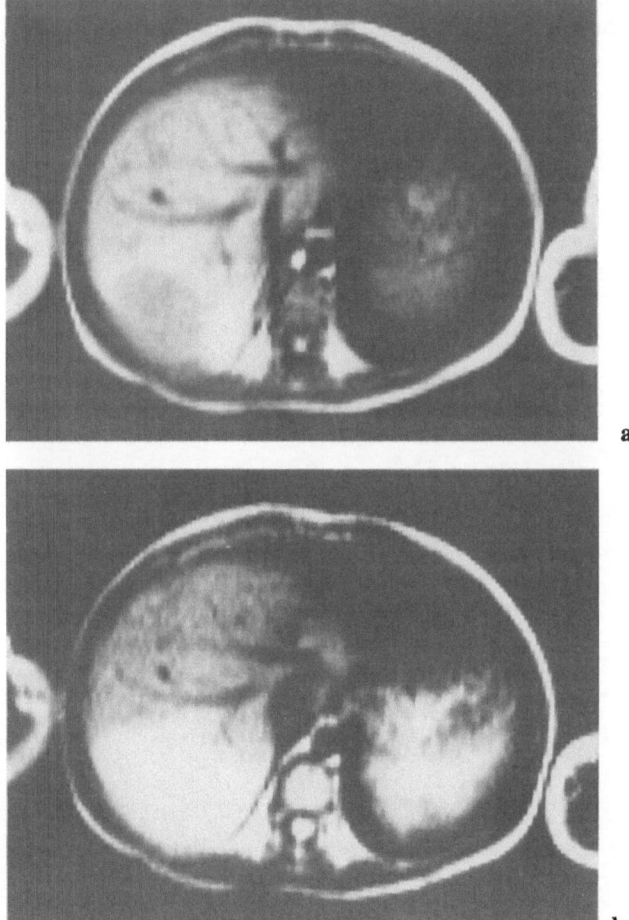

a

b

Fig. 11.10. **a** Transverse MR image in a child with liver abscess. SE 30/500 sequence reveals an area of decreased signal intensity posteriorly in the right lobe. The normal vascular structures are seen in the rest of the liver. **b** The T_2 weighted image performed in the same area using SE 60/1000 sequence demonstrates marked increased signal in the right lobe of the liver. The contrast between the abnormality and the normal liver is significantly greater on the T_2 weighted image.

the use of other imaging modalities together with clinical data (Fig. 11.11). Primary and metastatic neoplasms can be detected with NMR although optimal pulse sequences in diagnosing intrahepatic metastases are yet to be determined. The T_1 weighted images, as well as heavily T_2 weighted images, have improved soft tissue contrast between pathological and normal liver tissue when compared to computed tomography. In addition, the vascular structures within the liver are identified without the use of an intravenous contrast agent. Dilated biliary ducts are also differentiated from vascular structure on T_2 weighted images, since its prolonged T_2 relaxation times demonstrate increased signal intensity on these images, and this is coi.trasted well with markedly diminished signal from the vascular channels.

Nuclear magnetic resonance imaging is useful in patients with diffuse liver disease. Abnormally short T_1 values within the liver in patients with primary biliary cirrhosis may be related to high copper levels. Extensive iron deposits in the liver result in decreased T_1 and T_2 relaxation times due to the paramagnetic effect of the ferric ion. Similarly, excessive levels of copper within the liver will also decrease the T_1

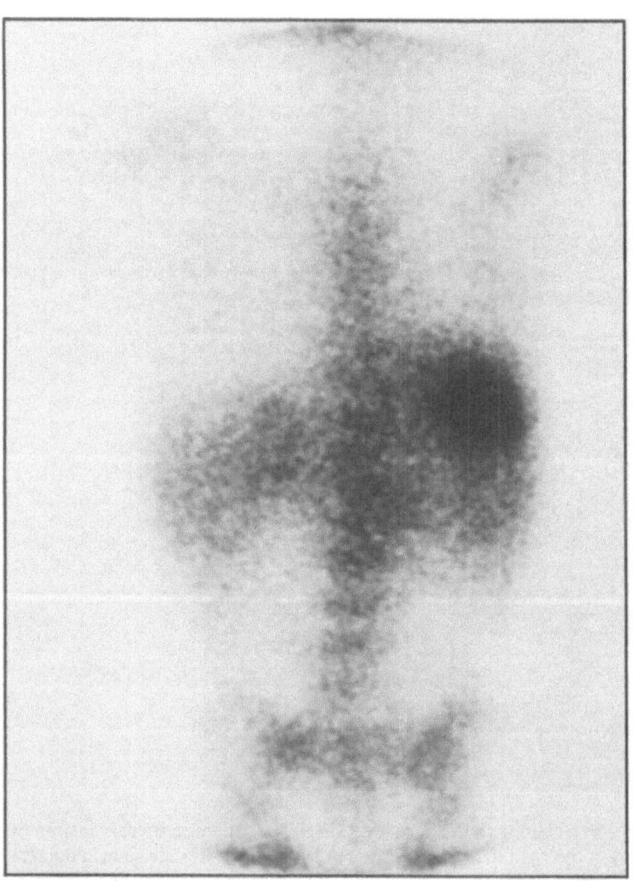

Fig. 11.11.   Focal area in the right lobe of the liver in the posterior view has increased activity on the Ga-67 citrate scan. Abscess has increased activity compared to normal liver uptake of Ga-67 citrate.

relaxation time. This could prove valuable in diagnosing patients with Wilson's disease. Cirrhosis of the liver has also been shown to have prolonged T_1 relaxation time.

Renal morphology is well evaluated by NMR since the intense signal from fat in the perinephric space provides good contrast. Simple cysts can be diagnosed because of their small borders and prolonged T_1 and T_2 relaxation values. Magnetic resonance is also used for the examination of primary as well as metastatic tumour in the kidney. Perinephric extension of the primary tumour in perinephric space can be identified with NMR. Metastatic spread from the renal tumour to the lymph node, the renal vein, and inferior vena cava, as well as bony metastases, can be detected by NMR, as by CT. However, unlike CT, no intravenous contrast administration is required with NMR.

Although NMR studies can demonstrate abnormalities in the kidney and metastatic spread to the bones, NMR is extremely useful to evaluate spread of the tumour to perinephric space, liver, lymph nodes and vascular invasion.

The retroperitoneum and the vascular structures of that area are well visualised using NMR. Lymph node involvement in the retroperitoneum necessary for staging of malignant diseases can also be evaluated by NMR. Patients with abnormality in the retroperitoneum shown by using Ga-67 citrate scanning can be studied with NMR to demonstrate individual groups of lymph node enlargement.

Skeleton

Although cortical bone is recorded on NMR images as an area of markedly decreased signal, the bone marrow produces an intense signal. Avascular necrosis of the femur can be diagnosed at an earlier stage on NMR than on conventional radiographs. A patient with avascular necrosis revealed decreased signal intensity in both femoral heads (SE 30/2000 pulse sequence) (Fig. 11.12). The radionuclide bone scan demonstrated increased activity in both femoral heads and a radiograph of the hips was normal. Our initial experience indicates that the radionuclide bone scan may be more sensitive than NMR in the diagnosis of avascular necrosis; this remains unsettled.

Although NMR has shown increased sensitivity in detecting skeletal pathology, its ability to differentiate neoplastic lesions from inflammatory or traumatic lesions is yet to be confirmed. In paediatric patients, NMR has demonstrated earlier detection of osteomyelitis than is currently possible radiographically (Fletcher et al. 1984). Nuclear medicine has shown an increased sensitivity to detection of osteomyelitis when compared to conventional radiographic procedures. Preliminary work in animals shows early detection of staphylococcus induced osteomyelitis with NMR (unpublished data). The exact role of nuclear medicine versus NMR in the diagnosis of osteomyelitis needs further study. Both benign and malignant primary osseous tumours as well as metastatic disease and inflammatory lesions have shown prolonged T_1 and T_2 relaxation times. In vivo measurements of T_1 relaxation time in pathological lesions have not been able to differentiate inflammatory lesions from neoplastic lesions at present.

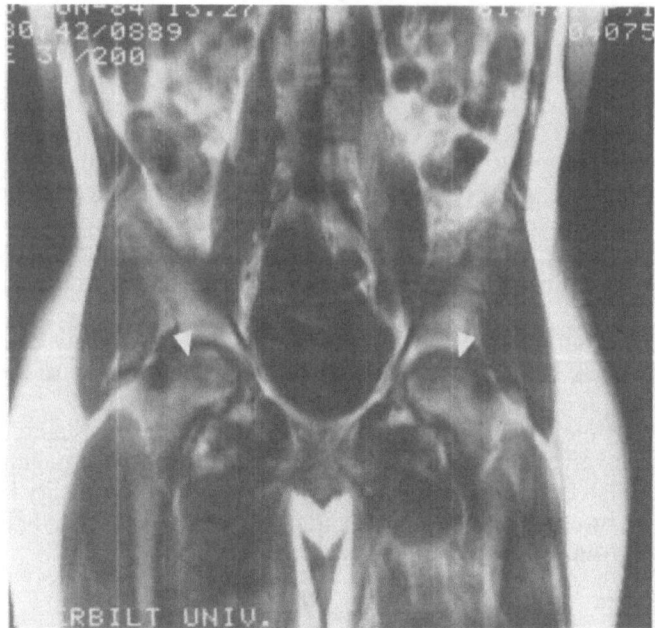

Fig. 11.12.   Coronal NMR image using 30/2000 pulse sequence demonstrates signal intensity in both femoral heads (*white arrowheads*) in a patient with bilateral avascular necrosis.

Conclusion

In summary, NMR imaging is a very useful technique in the evaluation of intracranial, cervical, thoracic, abdominal and skeletal pathology. Although tissue characterisation by NMR on the basis of T_1 and T_2 relaxation parameters alone is in the early stages of development, the use of in vivo spectroscopy will improve NMR diagnostic capabilities. The ability to perform imaging with C-13, P-31 and Na-23 will generate additional biochemical and physiological information. Oral and organ targeted contrast agents will further aid in both static and dynamic images. The use of surface coils will improve both spatial resolution and lesion detectability in small organs such as the thyroid and parathyroid glands. Evaluation of blood flow by NMR images using different pulse sequences, as well as gated cardiac images in cine format, will add to functional information similar to that obtained by nuclear medicine examinations. The role of NMR as a primary investigational modality and its complementary role with nuclear medicine and other imaging modalities requires further evaluation with larger series of cases for individual disease processes.

References

Araki T, Inouye T, Suzuki H, Machida T (1984) Magnetic resonance imaging of brain tumours: measurement of T_1. Radiology 150: 95–98
Basarab RM, Manni A, Harrison TS (1985) Dual isotope subtraction parathyroid scintigraphy in the preoperative evaluation of suspected hyperparathyroidism. Clin Nucl Med 10: 100

Bydder GM, Steiner RE, Young IR et al. (1982) Clinical NMR imaging of the brain: 140 cases. AJR 139: 215–236

Cohen AM, Criviston S, LiPuma JP, Bryan PJ, Haaga JR, Alfidi RJ (1983) NMR evaluation of hilar and mediastinal lymphadenopathy. Radiology 148: 739–742

de Certaines J, Herry JY, Lancien G, Benoist L, Bernard AM, Le Clech G (1982) Evaluation of human thyroid tumours by proton nuclear magnetic resonance. J Nucl Med 23: 48–51

Doyle FH, Pennock JM, Banis LM et al. (1982) Nuclear magnetic resonance imaging of the liver: initial experience. AJR 138: 193–200

Ferlin G, Borsato N, Camerani M et al. (1983) New perspectives in localizing enlarged parathyroids by technetium–thallium subtraction scan. J Nucl Med 24: 438

Fernandez-Ulloa M, Maxon HR, Mehta S, Sholiton LJ (1976) Iodine 131 uptake by primary lung adenocarcinoma: Misinterpretation of I-131 scan. JAMA 236: 857–858

Fletcher BD, Jacobstein MD, Nelson AD, Reimenschneider TA, Alfidi RJ (1984a) Gated magnetic resonance imaging of congenital cardiac malformation. Radiology 150: 137–140

Fletcher BD, Scoles PV, Nelson AD (1984b) Osteomyelitis in children: Detection by magnetic resonance. Radiology 150: 57–60

Lanzer P, Botvinick EH, Schiller NB et al. (1984) Cardiac imaging using gated magnetic resonance. Radiology 150: 121–127

Margulis AR, Higgins CB, Kaufman L, Crooks LE (1983) Clinical magnetic resonance imaging. San Francisco Radiology Research Education Foundation, San Francisco

Runge VM, Clanton JA, Price AC et al. (1984) Paramagnetic contrast agents in magnetic resonance imaging: Research at Vanderbilt University. Physiol Chem Phys Med NMR 16: 113–122

Sandler MP, Patton JA, Sacks GA, Shaff MI, Kulkarni MV, Partain CL (1984) Evaluation of intra-thoracic goiter with I-123 scintigraphy and nuclear magnetic resonance imaging. J Nucl Med 25: 847–876

Sandler MP, Patton JA, Partain CL (eds) (to be published) Thyroid and parathyroid imaging. Appleton-Century-Crofts, New York

Young AE, Gaunt JI, Croft DN et al. (1983) Location of parathyroid adenomas by thallium-201 and technetium-99m subtraction scanning. Br Med J 286: 1384

Subject Index